The History of the History of Mathematics

The History of the History of Mathematics

Case Studies for the Seventeenth, Eighteenth and Nineteenth Centuries

Edited by
Benjamin Wardhaugh

PETER LANG
Oxford • Bern • Berlin • Bruxelles • Frankfurt am Main • New York • Wien

Bibliographic information published by Die Deutsche Nationalbibliothek
Die Deutsche Nationalbibliothek lists this publication in the Deutsche Nationalbibliografie;
detailed bibliographic data is available on the Internet at http://dnb.d-nb.de.

A catalogue record for this book is available from the British Library.

Library of Congress Cataloging-in-Publication Data:

Wardhaugh, Benjamin, 1979-
 The history of the history of mathematics : case studies for the
seventeenth, eighteenth, and nineteenth centuries / Benjamin Wardhaugh.
   p. cm.
 Includes bibliographical references and index.
 ISBN 978-3-0343-0708-6 (alk. paper)
  1. Mathematics--Historiography--Case studies. 2. Mathematics
historians--Europe. I. Title.
 QA21.W3238 2012
 510.72'2--dc23
                    2011045102

All translations from quoted matter in the original language are by the author,
unless otherwise stated.

ISBN 978-3-0343-0708-6

Peter Lang AG, International Academic Publishers, Bern 2012
Hochfeldstrasse 32, CH-3012 Bern, Switzerland
info@peterlang.com, www.peterlang.com, www.peterlang.net

All rights reserved.
All parts of this publication are protected by copyright.
Any utilisation outside the strict limits of the copyright law, without the
permission of the publisher, is forbidden and liable to prosecution.
This applies in particular to reproductions, translations, microfilming,
and storage and processing in electronic retrieval systems.

Printed in Germany

Contents

Introduction   1

PHILIP BEELEY
The progress of Mathematick Learning:
John Wallis as historian of mathematics   9

BENJAMIN WARDHAUGH
'It must have commenced with mankind':
some ancient histories of arithmetic in eighteenth-century Britain   31

REBEKAH HIGGITT
The 'epitome of intellectual sagacity':
Biographical treatments of Newton as a mathematician   47

NICCOLÒ GUICCIARDINI
The Quarrel on the Invention of the Calculus
in Jean E. Montucla and Joseph Jérôme de Lalande,
Histoire des Mathématiques (1758/1799–1802)   73

ADRIAN RICE
Vindicating Leibniz in the calculus priority dispute:
The role of Augustus De Morgan   89

HENRIK KRAGH SØRENSEN
Reading Mittag-Leffler's biography of Abel
as an act of mathematical self-fashioning   115

JACQUELINE STEDALL
Thomas Harriot (1560–1621):
history and historiography   145

Bibliography   165

Notes on Contributors   181

Index   183

Introduction

> [O]ur history will embrace *all* mathematicians [...]. And not, moreover, in just a historical fashion – what age they lived in, what manner of life they led, what country they inhabited – but rather mathematically: what they wrote in what field, how well they wrote it and how useful it is for teaching beginners. Since I intended to say this, I could not, without fault, omit a discussion of the whole of mathematics and each of its branches.[1]

Mathematical histories have been written in Europe since the sixteenth century, yet on the whole there has been relatively little reflection on the trajectory which the history of mathematics itself has taken over time. Nor has sustained attention often been given to the historiography of a subject which by its nature involves methodological choices and dilemmas different from those of other kinds of history.[2] Henry Savile's demanding programme for the study of the history of mathematics, set out during his 1570 lectures on Ptolemy at Oxford and quoted above, illustrates the magnitude of the task facing the historian of mathematics. It also illustrates the tendency of mathematical histories to be dependent on particular understandings of the nature of mathematics, and of course to respond to the needs of particular audiences.

---

1   Oxford, Bodleian Library, MS Savile 29, fols 17^r–17^v, quoted and translated in Robert Goulding, *Defending Hypatia: Ramus, Savile, and the Renaissance rediscovery of mathematical history* (Dordrecht: Springer, 2010), 97.
2   Notable exceptions are Joseph W. Dauben and Christoph J. Scriba, eds, *Writing the history of mathematics: Its historical development* (Basel: Birkhäuser, 2002) and Amy Shell-Gellasch, 'Introduction: The Birth and Growth of a Community' and Ivor Grattan-Guinness, 'History or Heritage? An Important Distinction in Mathematics and for Mathematics Education', both in Glen Van Brummelen and Michael Kinyon, eds, *Mathematics and the Historian's Craft: The Kenneth O. May Lectures* (New York: Springer, 2005), 3–6 and 7–22.

The history of mathematics in the sixteenth century has been addressed by Robert Goulding in his recent book on Peter Ramus and Henry Savile.[3] Ramus's *Prooemium mathematicum* of 1567 was one of the earliest published histories of ancient mathematics. Widely read at the time, it continued to influence mathematicians and writers on the history of mathematics over the next century, partly because of its comprehensive scope, but equally because of the practical and progressive lens through which Ramus observed, selected and arranged his sources on the history of mathematics. One of the earliest extended critical responses to Ramus's history is found in Savile's lectures on Ptolemy. There, in the early part of those lectures, Savile presented a history of ancient mathematics based almost entirely on his study of Ramus's *Prooemium*, but arguing for an entirely opposite account of mathematics: not practical and changeable, but theoretical and eternal. What Goulding has called the 'malleability' of the evidence available to Renaissance scholars concerning the history of ancient mathematics was manifested most dramatically in the two men's divergent attitudes to that most famous of mathematical texts, Euclid's *Elements*. Where Savile saw a single 'most beautiful body', Ramus wished to 'pull apart the bones, flesh, spirit, and blood' in an attempt to 'cure the disease' he found in a flawed and corrupt text.[4] Thus divergent attitudes to mathematics could lead to radically different textual practices, to entirely opposed understandings of the mathematical past as history, and to wholesale disagreement concerning the interpretation of historical mathematicians and their work.

The mathematical narratives of Ramus and Savile set the stage for the later development of mathematical history writing. Later historians would face some of the same issues and replay some of the same types of disagreement in respect – often – not of ancient but of modern mathematics. Like Ramus and Savile, they would be concerned not just to construct but to use the mathematical past, their agendas shaped by national and local

3   Goulding, *Defending Hypatia*.
4   Henry Savile, *Praelectiones tresdecim in principium Elementorum Euclidis Oxonii habitae MDCXX* (Oxford: Iohannes Lichfield, & Iacobus Short, 1621), 140; Petrus Ramus, *Scholarum mathematicarum libri unus et triginta* (Basel: Episcopius, 1569), 91, trans. in Goulding 177, 170 respectively.

considerations as well as by differing assumptions about the nature of mathematics and mathematicians. This volume presents seven case studies illustrating the diversity which resulted in thinking and writing about the mathematical past from the early modern period until the early twentieth century.

During the second half of the seventeenth century, the growth of scientific communication contributed to major advances in mathematical knowledge, but it also engendered an increasingly bitter spirit of competition, expressed in the numerous disputes over priority in discovery which plagued the Republic of Letters. History of mathematics could effectively become a cover for establishing a certain author's claim to priority, as exemplified for instance by the historical accounts of the cycloid produced variously by Blaise Pascal, Carlo Dati, and Johann Gröning, and it was all too often a self-serving enterprise rather than anything more.

John Wallis, the Savilian professor of geometry at the University of Oxford, was not completely averse to this new kind of historical writing. But in his *Treatise of Algebra* he embarked on a much broader historical mission, seeking to evince the ancient roots of algebra and to show how it had progressed through the centuries to the heights it had attained in his day. His project was arguably a *historia* in an Aristotelian sense, concerned to document facts rather than to discover causes. The results were not entirely free of the biases of party and nation, but Wallis's conception of history was neither unsophisticated nor inherently one-sided. By putting the *Treatise of Algebra* in its scientific and cultural setting, Philip Beeley attempts to resolve the evident tension in Wallis's work between different types of concerns, and thus to reassess his legacy as a historian of mathematics.

By the eighteenth century, an interest in the ancient mathematical past – already evinced by Ramus and Savile, and by Wallis and his contemporaries – was beginning to find a place even in the most popular accounts of mathematical subjects such as arithmetic primers and dictionaries, with consequences for the way mathematical history, and therefore mathematics, were presented to unsophisticated readers. If learners of arithmetic were to be motivated and encouraged they should ideally be presented with a convincing ancient pedigree for their subject: yet the available historical sources for the earliest mathematics hardly enabled one to be constructed. Writers fell back on unabashed speculation or added a Christianizing spin to a small selection of ancient materials, gravely suggesting even that 'some Method of Numbering was used by *Adam* and *Eve* in *Paradise*', and thereby writing mathematics into history in ways previously unthought of. Benjamin Wardhaugh's chapter considers these popular accounts of the origins of arithmetic written in eighteenth-century England, and asks what they tell us about the developing reputation of mathematics and its history.

Equally important for that reputation, and for the developing genres of mathematical history-writing and of mathematical biography, one of the defining issues in the eighteenth and nineteenth centuries was the treatment of specific prominent mathematicians of the recent past. This was true of no-one more than of Isaac Newton and Gottfried Wilhelm Leibniz. Newton's reputation would come to tower over British science and mathematics, and his quarrel with Leibniz was a locus for a remarkable quantity of historical and biographical assessment. Three chapters in this volume examine different ways in which writings about Newton and about the Newton–Leibniz dispute illuminate the development of mathematical history and biography, from the eighteenth to the twentieth century.

Rebekah Higgitt considers the depiction of Isaac Newton as a mathematician in biographies across that period. As with other aspects of Newton's life and work, the discussion of his mathematics varied over time as views of the discipline and its practitioners underwent significant change. At the same time, national context and disciplinary and personal interests all played roles in shifting perceptions of Newton's life, personality and work, and the relationship between them. While Newton's mathematical accomplishments continued to be revered, there was some criticism, even in biography, of the obscurity of his published work. This issue was particularly important at key periods, such as when the distinctive Continental and British traditions were established in the early eighteenth century, and when they were largely reunited a century later. Alongside such concerns we also find more popular portrayals that largely avoid detailed consideration of Newton's mathematics, effectively sidelining what many considered Newton's most significant work, or contributing to a popular image of the mathematical genius. Tracking how Newton the mathematician has, or has not, been integrated with dominant themes in Newtonian biography not only illustrates the history of mathematical history; it also provides a window onto changing views about the relationship between mathematics and other branches of science, and the role of mathematics in considering the persona of the man of science.

A work which dealt with the calculus controversy between Newton and Leibniz in particular detail was Jean E. Montucla's *Histoire des Mathématiques*, first published in 1758 and revised with the contribution of Joseph Jérôme de Lalande around the turn of the century. Niccolò Guicciardini considers the image of the calculus controversy conveyed in this monumental history, and draws comparisons with contemporary British historical work, including that of Hutton, Rigaud, and Brewster. He shows how these diverse accounts of the notorious controversy reflect the diverse agendas of the historians concerned. Montucla's was not a nationalistic account, but a balanced one in which the calculus was conceived as emerging from the contributions of many individuals over an extended period. It was shaped by the milieu of the French encyclopedists, for whom history was expected to show the progress of knowledge as a matter of universal, enlightened, cooperation.

Thus differences in national context and intellectual agenda could result not just in different judgements about the narrow issue of Newton *vs*. Leibniz, but also in different understandings of what it might mean to 'invent the calculus', and of what criteria should properly be used to assess matters of intellectual priority and discovery. But despite the existence in print of such sophisticated assessments as Montucla's, British mathematicians in the nineteenth century continued to regard Leibniz as an underhanded plagiarist, an attitude reinforced by the virtual deification of Newton by his British biographers. One of the first to question this view was the nineteenth-century mathematician Augustus De Morgan, who, in a series of works published between 1846 and 1855, attempted to set the historical record straight. Adrian Rice examines De Morgan's research in this area and investigates the motivations that led him to initiate the rehabilitation of Leibniz among British mathematicians. His position as a religious nonconformist and his critical stance towards both the Church of

England and the Royal Society produced a readiness to take the part of the underdog and to – in Higgitt's phrase – pay 'attention to the little men of mathematical history' (a position which converges in some ways with that of Montucla, although emerging from a very different milieu). Although his instincts led De Morgan astray in the notorious case of the scholar/ thief Guglielmo Libri, his historical assessment of Newton and Leibniz would come to be widely accepted among British historians of mathematics, changing permanently the perception of both men in Britain.

The changing reputation of Newton, and the impact on mathematical history of nineteenth-century ideas about the nature of intellectual achievement, may be compared with the case discussed by Henrik Kragh Sørensen. The Norwegian mathematician Niels Henrik Abel lived a life that could be called worthy of a romantic hero, neglected and defiant against the background of Norway's struggle for independent nationhood. He was born in 1802, when Norway formed part of a union with Denmark; by the time of his education at the University in Christiania (now Oslo) it had entered into a union with Sweden after a brief spell of independence. Abel's mathematical research, published mainly in Berlin, became internationally renowned, yet he died in 1829, before the Norwegian state (or any other state, for that matter) could provide him with a permanent income. What were immediate, practical issues for Abel became symbols to later generations of Norwegian mathematicians, who used his legacy to position themselves in debates within mathematics and beyond. That legacy developed in various ways in the process.

The celebrations in Christiania for the 1902 centenary of Abel's birth took on a political aspect in relation to the union between Norway and Sweden. The Swedish mathematical entrepreneur Gösta Mittag-Leffler, an admirer of Abel and an adept self-fashioner, seized the opportunity of the celebrations to promote himself and his journal *Acta Mathematica*. Later, during the period of political tension when the union was moving towards its eventual dissolution in 1905, he published his own biographical account of Abel. That biography was both a piece of self-fashioning, a defence of a particular style of mathematical research, and a contribution to contemporary political debates. Thus we once again see the history of mathematics both shaping, and being shaped by, the events which surround it, as well as interacting with specific national and individual attitudes to mathematics and to mathematicians.

Light is shed on these issues, and on how they are being transformed by technological developments in the twenty-first century, by Jacqueline Stedall's discussion of Thomas Harriot and his editors and historians, a case which spans the whole historical period covered by this book. Harriot's reputation has risen and fallen over the last four centuries: contemporaries and immediate successors admired him, but the posthumous publication of some his work in 1631 obscured the exact nature of his achievements, and searches for his papers later in the seventeenth century drew a blank. Meanwhile John Wallis's history of algebra included controversial claims that Descartes had plagiarized from Harriot. By the time Harriot's papers were rediscovered in the late eighteenth century their historical context was little understood, while a project to publish the papers foundered, defeated by their complexity and disorder. Harriot became a figure of myth in popular history, a position which continued into the late twentieth century. A number of recent publications have now put parts of the Harriot papers into print, and scholarship has begun to recover his achievements. Stedall considers the different approaches that have been taken to Harriot, and wonders whether the current project to digitize his papers will, too, come to be seen as a product of its time and place.

The papers in this volume take the story of the history of mathematics up to the early twentieth century; certain of them also examine more recent developments, and Stedall's considers the transformation of editorial practice which is taking place at the present day. During the last hundred or hundred and fifty years one of the most important processes for the construction of the mathematical past and of the reputations of individual historical mathematicians has been the production of critical editions, in a process analogous to the 'canonisation' of early modern and literary figures by the application to their works of the meticulous editorial procedures originally developed for work on classical texts. This phenomenon has not yet received the critical reflection which it seems to deserve. How, and to what degree, is the mathematical case special compared with the production of critical editions in other fields? What are the implications for the ways – and the circumstances – in which critical editions of historical mathematics can be created and used, and therefore for the construction of mathematical history on their basis?

At the same time, the audiences for historical mathematics continue to change. In the last half-century national societies for the history of mathematics have been founded in several countries, while the volume and range of writing being produced about the subject have grown considerably. New agendas influence research on the history of mathematics: notably the role of history in the teaching of mathematics at various levels. Debate continues about the value and appropriateness of different approaches to historical mathematical texts, approaches which are intended not just to form and fix the reputations of the authors and the works they concern, but also to speak to different audiences and produce histories or editions suitable for different uses: pedagaogy, the popularization and communication of mathematics, or shaping the reputation of mathematics itself, say. The characteristic concerns of the history of mathematics have thus perhaps shifted somewhat away from considerations of national and personal reputations, and somewhat towards the merits of particular textual practices and their suitability for particular situations.

As this book goes to press a conference on mathematical editing since 1900 is in preparation in Oxford, and it is to be hoped that – as new technologies and new audiences continue to tranform textual practices in every discipline – this will provide a stimulus for reflections which will help to understand the history of the history of mathematics in the twentieth century.

The editor wishes to take this opportunity to thank all of the contributors to this volume, as well as all of the participants at the conference on 'The History of the History of Mathematics' which took place in Oxford in December 2010. He is grateful to All Souls College, which hosted that conference, and to the British Society for the History of Mathematics, the International Commission on the History of Mathematics, the History Faculty at the University of Oxford, and All Souls College, which supported it financially.

PHILIP BEELEY
LINACRE COLLEGE, UNIVERSITY OF OXFORD

The progress of Mathematick Learning: John Wallis as historian of mathematics

Introduction

Of all the relics the learned and great of former times have handed down to posterity pictorial images are sometimes the most revealing. When John Wallis (1616–1703) sat for the court and society painter Godfrey Kneller in 1701 (see Figure 1), he made sure that a small table in the background was decorated with two of the most powerful symbols of his intellectual prowess: the gold chain awarded to him in 1691 by Frederick III, Elector of Brandenburg for his services in deciphering intercepted letters, and the second volume of his *Opera mathematica*, printed at the Sheldonian Theatre in 1693 with the title 'Algebra' emblazoned on its spine. This was after all a portrait of which he was to be especially proud. Commissioned by his friend Samuel Pepys, it was to be presented to the University of Oxford as a lasting memorial to one of its most renowned scholars who had served as Savilian professor of geometry for over fifty years.

Of the two objects on the table one made perfect sense. As a decipherer, Wallis's achievements stood unparallelled in the second half of the seventeenth century and he was justly seen as Europe's greatest exponent of the art.[1] But it is perhaps less clear why of all his numerous mathematical works

---

[1] Gottfried Wilhelm Leibniz, ed. Prussian Academy of Sciences (and successors), *Sämtliche Schriften und Briefe* (Darmstadt: Reichl Verlag (and successors), 1923–), Reihe I, vol. 16, p. 726; Philip Beeley, '"Un de mes amis". On Leibniz's relation to the English mathematician and theologian John Wallis', in Pauline Phemister and Stuart

Figure 1: John Wallis, by Godfrey Kneller, 1701.

he chose the *Algebra* where many, even Wallis himself, considered earlier books such as *De sectionibus conicis* (1655), the *Arithmetica infinitorum* (1656) or the *Mechanica: sive, de motu, tractatus geometricus* (1670–71), to have been his most significant contributions to the growth of the mathematical sciences.

Wallis's *Algebra*, or, to give the complete title, *A Treatise of Algebra, both Historical and Practical*, is a book whose own history is worthy of consideration. Ostensibly an 'Account of the Original, Progress, and Advancement of (what we now call) Algebra, from time to time; shewing its true Antiquity (as far as I have been able to trace it;) and by what Steps it hath attained

---

Brown, eds, *Leibniz and the English-Speaking World* (Dordrecht: Springer, 2007), 63–81; David E. Smith, 'John Wallis as a Cryptographer', *Bulletin of the American Mathematical Society* 24 (1917), 83–96 and 166–9.

to the Height at which now it is,'[2] the first edition appeared in English in 1685. However, much of the book is considerably older. A substantial part of the hundred chapters (in the second Latin edition comprising most of volume two of his *Opera mathematica* the number increases to one hundred and twelve) was already completed by April 1677, when 'A large discourse concerning algebra' was one of six works which Wallis sent to the mathematical intelligencer John Collins in London with a prospect to being published.[3]

In the following, we consider the *Treatise of Algebra* under three historical perspectives. First, we attempt to reconstruct the history of the book itself from manuscript and printed sources, including Wallis's extensive scientific correspondence. Second, we proceed from here to an historical account of the context in which the *Treatise of Algebra* should be understood. Particular emphasis is given thereby to the expectations of the Royal Society, which in 1683 agreed to underwrite the publication. Third, we seek to elucidate the sense in which Wallis describes his *Treatise of Algebra* as being 'both historical and practical'. In this way, we seek to assess not only the character of Wallis's historical approach in general, but also Wallis's character as a historian of mathematics, if indeed such a description is appropriate.

Finding a publisher for the *Treatise of Algebra*

Although Wallis sent Collins his manuscript discourse on algebra for possible publication in 1677, his intention to produce a textbook on algebra was in fact much older. At the end of his *Mathesis universalis*, or as he preferred

---

[2] John Wallis, '[Account of] A Treatise of Algebra, both historical and practical', *Philosophical Transactions* 173 (22 July 1685), 1095–1105, at 1095.

[3] Stephen Peter Rigaud and Stephen Jordan Rigaud, eds, *Correspondence of Scientific Men of the Seventeenth Century* (2 vols, Oxford: Oxford University Press, 1841), vol. 2, p. 607.

to call it, his *Opus arithmeticum*, which grew out of his Oxford lectures and was published in 1657, the Savilian professor pointed out that he had wanted to add something on the doctrine of analysis, but in view of the size of what he had already drafted on the topic he felt it would require a separate volume instead.[4] It is therefore probable that at least the first draft of either part or the whole of what he sent to Collins twenty years later originated from that time.

We do not know why twenty years elapsed before the first steps were taken to realize the intention expressed in *Mathesis universalis*. The most likely explanation is that Wallis tried to find someone in the printing trade willing to take on the work during these two decades, but was unable to meet with success. This would also explain why Collins, who like no other knew the ins and outs of the London mathematical book trade, eventually came to present the only solution. From letters exchanged between Wallis and the man described as the English Mersenne by Issac Barrow[5] we know that Collins offered to serve as Wallis's agent in helping to bring about the publication of his mathematical texts. Wallis was not the only English mathematician who would benefit from Collins's efforts in this regard, but certainly the most illustrious. Shortly before his death in 1683, Wallis noted in a letter to John Aubrey that 'the progress of Mathematick Learning' owed much to Collins's industry.[6]

Collins's task was not an easy one. In contrast to the Low Countries and to Italy, England lacked a strong and established market in mathematical books. Only a handful of the large number of printers in London were prepared to take on the financial risks in producing such books – elementary works excepted – and authors could therefore easily find themselves in the position of the Savilian professor. Sometimes combining more practical

---

[4] John Wallis, *Mathesis universalis: sive, arithmeticum opus integrum* (Oxford: Leonard Lichfield for Thomas Robinson, 1657), 398; John Wallis, *Opera mathematica* (3 vols, Oxford: at the Sheldonian Theatre, 1693–9), vol. 1, p. 228.

[5] Isaac Barrow, *Lectiones XVIII, Cantabrigiae in Scholis publicis habitae; in quibus opticorum phaenomenwn genuinae rationes investigantur, ac exponuntur* (London: W. Godbid, 1669), 6.

[6] Oxford, Bodleian Library, MS Aubrey 13, fol. 243^r.

works with more theoretical provided a solution and Wallis alludes to this possibility in his letter to Collins of April 1677: 'What you think fit of these to put to the press, either by themselves or with the books you mention, you may freely do it; only I would be content to review it before it be printed'.[7] The language of the publication was also an important factor in getting it accepted. Already two months earlier, Wallis had offered if necessary to translate into English those works which he had submitted in Latin.[8]

No records survive of Collins's efforts to persuade local printers to take on the production of the most substantial of the works submitted by Wallis over the course of the following months. It would have been apparent that the typesetting alone involved considerable time and cost and we can only assume that for this reason they were unsuccessful. However, another possibility arose. By the beginning of 1680 Collins had landed on the idea of persuading the Royal Society to underwrite the printing of 'two volumes of algebra, written by Dr Wallis, Mr Baker, Mr Newton, &c'. At the meeting of Council on 21 January, William Croone, one of the founding fellows, put forward this proposal, the core of which was that while Collins would arrange the printing, the Society would 'engage to take off 60 copies after the rate of 1d½ a sheet'.[9] No doubt Newton's notorious reluctance to publish prematurely precluded his participation, while Thomas Baker's *Geometrical Key, or, Gate of Equations Unlocked* was slow in production – although the Royal Society did approve its publication in 1682.

The written proposal about printing the two volumes, which Council had demanded from Collins as an understandable prerequisite before making any decision, eventually bore the title 'A proposal about printing a treatise of algebra, historical and practical' and named John Wallis as its sole author. Abraham Hill, another founding fellow of the Royal Society and that institution's long-serving treasurer, presented the proposal under the new conditions of sixty copies at 'three half pence a sheet' to Council

---

7   Rigaud, *Correspondence of Scientific Men*, vol. 2, p. 607.
8   Ibid., 605.
9   Thomas Birch, *The History of the Royal Society of London for Improving of Natural Knowledge* (4 vols, London: for A. Millar, 1756–7), vol. 4, p. 4.

on 22 November 1682, whereupon it was duly accepted.[10] The proposal explicitly refers to the Treatise as having been written in 1676 and since then enlarged 'so as to contain not only an history, but likewise an institution of algebra, according to the several methods hitherto in practice, with many additions of his own.' Robert Hooke subsequently communicated the contents of the proposal to the members of the Society present, the secretary suggesting that if printed the book would comprise around fifty or sixty books in quires. Since the guarantee that the Society would take sixty copies was not deemed sufficient for the prospective printer John Playford, subscriptions were also sought. Playford, a specialist in printing music, who alone possessed the necessary range of type, was evidently aware of the limited market in mathematical books. In order to reach a wider circle of potential subscribers, the Oxford bookseller Richard Davis arranged for the written proposal to be printed separately. The proposal informed its mathematically interested readers of the local booksellers they needed to contact in order to be sure of obtaining a copy.

The Proposal

Although ostensibly from Collins, the Proposal was in fact produced by Wallis himself. In it, the Savilian professor sets out his admittedly largely already realized plan to begin by showing that algebra was already in use among the ancient Greeks 'but studiously concealed as a great Secret'.[11] From discussions of early examples in Euclid (fl. 300 BC), Archimedes (c. 287 BC–c. 212 BC), Apollonius (c. 262–190 BC), Pappus (c. 290–350) and particularly Diophantus (c. 200–284) who produced the first known professed treatise of algebra, Wallis would describe its employment among

10   Ibid., 167.
11   John Wallis, *A Proposal about Printing a Treatise of Algebra, Historical and Practical* (London: John Playford, 1683), [1].

Arabic authors, and from there proceed to the first modern printed author Luca Pacioli (Lucas du Burgo) (1446/7–1517). The treatment of cubic equations in Scipione del Ferro (Scipio Ferrens) (1465–1526), Girolamo Cardano (1501–76), and Niccolò Fontana Tartaglia (1499/1500–57) would be considered just as that of Rafael Bombelli (1526–72), who proceeded 'yet farther' than his predecessors in showing 'how to reduce a Biquadratic Equations (by the help of a Cubic) to two Quadratics'.[12] From here he would progress to the specious arithmetic of François Viète (1540–1603) and its improvement in William Oughtred (1575–1660), excusing the rather full presentation of the latter 'though much had been taught in the Authors above-mentioned' due to its delivery in Oughtred 'in the most compendious form'.[13] Wallis also excuses his extensive treatment of Harriot, with whom a quarter of all the chapters is concerned, arguing that Warner's edition of the *Artis analyticae praxis* (1631) 'hath been but little known abroad' and that through his efforts 'it may hence appear to what estate Harriot had brought Algebra before his death'.[14] Moreover, Wallis uses the opportunity to assert that Harriot had taught all that

> which hath since passed for the Cartesian method of Algebra, there being scarce any thing of (pure) Algebra in Des Cartes, which was not before in Harriot, from whom Des Cartes seems to have taken what he hath, that is purely Algebra, but without naming him.[15]

Neither the suggestion of Descartes's plagiarism nor the promise of an account of Pell's (1611–85) method of equations would explain the inordinate interest which Wallis's *Treatise of Algebra* gained even beyond England's borders long before its publication. Rather, it was Wallis's declared intention to show, finally, how building on Cavalieri's (1598–1647) method of indivisibles and on his own arithmetic of infinities, Mercator (1620?–87)

---

12   Ibid., [2].
13   Ibid., [3].
14   Ibid.
15   Ibid. See Jackie Stedall's chapter in this volume for more about these claims concerning Harriot.

and Newton (1642–1727) through their work on series had succeeded in advancing tetragonistic methods beyond anything hitherto achieved.

Publication, promotion, and reception of the *Treatise of Algebra*

In view of this build-up, the manner in which the *Treatise of Algebra* finally entered the learned world was less than prodigious. Publication coincided with one of Wallis's visits to London, and he was able to present a copy in person during the meeting of the Royal Society he attended on 6 May 1685. But any joy he might have experienced at completing the project would have been tempered by the discovery of numerous printing errors – graphic proof of the difficulties of publishing mathematical books at a distance in the seventeenth century. In the copy Wallis gave to the Royal Society the faults were corrected with his own hand.[16] As if this were not bad enough, the printer delivered a message saying that the sixty underwritten copies would only be delivered to the Society against payment of sixty pounds. When no alternative financial agreement with Playford could be reached – the Royal Society was exceptionally short of money because of its commitment to publish John Ray's *Historia piscium* (1686) – Wallis himself stepped in and undertook to have twelve copies delivered. There is no record of this happening, but it was possibly to compensate for the costs which Wallis incurred in supplying copies of his *Treatise of Algebra* that his membership arrears were struck off in the autumn of the same year.[17]

Nor did the Royal Society's promotion of the *Treatise of Algebra* cease with the book's publication. Three months after it had appeared, the secretary of the Royal Society, Francis Aston, prevailed upon Wallis to publish

---

16 Birch, *The History of the Royal Society*, vol. 4, p. 396.
17 Ibid., 426.

a short account in the *Philosophical Transactions*.[18] Wallis obliged with an enlarged version of his text for the original proposal which had in the meantime already appeared again, somewhat enlarged, as the preface to the published book.[19] This all came at a time of heightened activity on the part of Wallis within the Royal Society, no doubt reflecting potentially dangerous moves against him back in Oxford. Following the ascent of the openly Roman Catholic James II to the throne in February 1685, Wallis's enemies in the University were evidently seeking actively to unseat him.[20]

None of the scientific journals which appeared in France, Italy or the Low Countries took notice of the *Treatise of Algebra*. In Germany things were rather different, where at Otto Mencke's request Leibniz reviewed the book in the June 1686 issue of the *Acta eruditorum*. Barely a word of criticism is to be found. Leibniz summarizes with equanimity Wallis's account of the incremental development of algebra since ancient times, adding some names which the author had neglected to mention and correcting some details on priority, most notably that Heuraet (1634–c. 1660) had succeeded in rectifying any curve before William Neile (1637–70) or Christopher Wren (1632–1723). Admittedly, he regards Oughtred's contribution as having been somewhat less than Wallis had proposed, but he concurs almost unrestrictedly with Wallis's presentation of Harriot, quoting with delight the famous story of Roberval's reaction at being shown Warner's edition of the *Artis analyticae praxis* by William Cavendish.[21] Indeed, Leibniz did not need much persuading of Descartes's supposed plagiarism having seen his literary remains while he was living in Paris in the 1670s.[22] Remarkably, he devotes just one line to Wallis's account of

---

18 Robert William Theodore Gunther, *Early Science in Oxford*, vol. 12: *Dr. Plot and the Correspondence of the Philosophical Society of Oxford* (Oxford: Oxford University Press, 1939), 100.
19 Wallis, '[Account of] A Treatise of Algebra'.
20 London, British Library, Add. MS. 32499, fol. 377^r.
21 Gottfried Wilhelm Leibniz, review of Wallis, *Treatise of Algebra*, *Acta eruditorum* (June 1686), 283–9, at 285–6.
22 Philip Beeley and Christoph J. Scriba, 'Wallis, Leibniz und der Fall von Harriot und Descartes. Zur Geschichte eines vermeintlichen Plagiats im 17. Jahrhundert', *Acta Historica Leopoldina* 45 (2005), 115–29.

Newton, compared to more than one complete page to the various tracts Wallis had appended as a means to illustrate the practical applications of algebra, including *Of Combinations, Alternations, and Aliquot Parts* and *Cono-Cuneus: Or, the Shipwright's Circular Wedge*. Some of these were evidently considerably older than the *Treatise of Algebra* itself.

Of course, when the second edition of the *Treatise of Algebra* appeared in 1693, the passages on Newton and Leibniz in chapter XCV had changed considerably. By then, the *Treatise of Algebra* had become instrumental in establishing Newton's claim to priority at Leibniz's expense. But that story does not concern us here.[23]

Designs of the Royal Society

Although Wallis's *Treatise of Algebra* was not in any sense commissioned – its origins in the late 1650s speak against this – it is important to take into account the possible impact that institutional considerations might have had on its final composition. The Royal Society, while originating from an ideal of scientific endeavour and collaboration across national boundaries so deeply rooted in the tradition of Hartlib, Comenius, and Oldenburg, had found itself since the late 1660s almost in constant need of self-justification, such as that provided by Sprat's rather Baconian *History*.[24] While it was not incumbent upon the Royal Society to publish books under its imprimatur in English or even to promote English authors, there was certainly a sense of its being proper that its resources should where possible be used in this way.

---

23 See the chapters by Rebekah Higgitt, Niccolò Guicciardini and Adrian Rice in this volume.

24 Thomas Sprat, *The History of the Royal-Society of London* (London: Thomas Roycroft for J. Martyn and J. Allestry, 1667).

The progress of Mathematick Learning 19

Thus when Wallis apologizes to Collins that the discourse on algebra which he had deposited with him was in English rather than in Latin, Collins points out 'that tis the Designe of the Royall Societie to promote and encourage the printing of Mathematicks and other books of Art in our own tounge.'[25] Collins also draws attention to Wallis's account of 'the learned paines of Englishmen', naming Harriot, Wallis himself, and Newton, as if to imply that this, too, belonged to the design of the Royal Society. Interestingly, he proceeds to recommend that Barrow be included – which indeed happened – before suggesting improvements to Wallis's presentation of the development of Newton's work on series.[26]

Such participation would clearly not have been unwelcome to Wallis, who had explicitly asked Collins for the names of some ancient mathematical authors he thought might be worth his considering. Collins's posthumously published letter to Wallis of 3 October 1682, on some defects in algebra, reads like an extensive list of remarks made by an avid reader of the text for inclusion in or the amendment of the *Treatise of Algebra*.[27] Indeed, there is evidence that right up to the publication corrections, emendations and inclusions were made, though clearly not enough to prevent embarrassing printing errors. Three years after publication of the *Treatise of Algebra*, Wallis wrote to the Oxford-educated diallist Thomas Strode to apologize for inserting a letter of his without first writing to seek permission: 'For the Book being then near finishing there was not time to write letters back & forth concerning it.'[28] He then employs a standard phrase with a view to his intended Latin edition: 'if you think fit to have any thing altered, added, or omitted, of what therein concerns yourself, I shall readily comply

---

25 Isaac Newton, ed. Herbert W. Turnbull, Joseph F. Scott, A. Rupert Hall and Laura Tilling, *The Correspondence of Isaac Newton* (7 vols, Cambridge: Cambridge University Press, 1967–77), vol. 2, p. 241.
26 Ibid., 242.
27 John Collins, 'A Letter from Mr. John Collins to the Reverend and Learned Dr. John Wallis Savilian Professor of Geometry in the University of Oxford, giving his thoughts about some Defects in Algebra', *Philosophical Transactions* 159 (20 May 1684), 575–82.
28 Oxford, Bodleian Library, MS Savile G 25 (6).

with you.' Wallis introduced the letter from Strode of 3 November 1684 on the geometric construction of quadratic equations in the extremely rich 'Additions and Emendations' section at the end of the *Treatise*, printing at the same time his own reply to that letter.[29]

In what sense a history?

To be sure, the mere presence of the word historical in the title of Wallis's *Treatise of Algebra* can lead to misunderstandings as to what he was setting out to achieve. Writing at a time when history was yet to become established as an academic discipline in its own right at either of the English universities, Wallis straddles at least two if not more historical traditions. Although history is not a topos in any of his extant letters or writings, and certainly not in the *Treatise of Algebra* itself, it is clear that many of his historical remarks are deeply rooted in the Aristotelian tradition which quite clearly distinguishes *historia* from *scientia* and the knowledge of causes.[30]

Histories in this Aristotelian sense were concerned with a body of individual facts rather than with causal developments. There was nothing innately chronological about history, and in the seventeenth century the atemporal Aristotelian concept comfortably sat alongside chronological accounts of church councils, wars, nations and the like.[31] Nor was it

---

29 John Wallis, *A Treatise of Algebra, both historical and practical* (London: John Playford for Richard Davis, 1685), 'Additions and Emendations', 162–6.
30 Per Landgren, *Det aristoteliska historiebegreppet. Historieteori i renässansens Europa och Sverige* (Gothenburg: Göteborgs Universitet, 2008), 46ff; cf. Edmund Halley, 'An Historical Account of the Trade Winds, and Monsoons, Observable in the Seas between and near the Tropicks, with an Attempt to Assign the Phisical Cause of the Said Winds', *Philosophical Transactions* 183 (July, August and September 1686), 153–68.
31 Landgren, *Det aristoteliska historiebegreppet*, 25ff; cf. 43ff.; Thierry Hoquet, 'History without Time: Buffon's Natural History as a Nonmathematical Physique', *Isis* 101 (2010), 30–61.

understood that in a historical presentation the evolution of ideas would be documented as a quasi-causal process in which an author would receive and transform concepts and results found in other sources. For Leibniz, the published book is to be seen as a monument of historical truth, precisely because it represents factual knowledge.[32] Likewise, Gerardus Vossius (1577–1649), Wallis's favourite source on historical questions in mathematics, emphasizes factual rather than causal considerations. Mirroring what Wallis at least in part sets out to achieve in his *Treatise of Algebra*, Vossius writes of *historia literaria*: 'The history of letters, or academic history, is about learned men, their writings and the steady growth of knowledge. It is also about practical discoveries and their advancement.'[33]

It is our contention that Wallis in his *Treatise of Algebra* seeks only to produce something broadly corresponding to the tradition of *historia literaria* and not a history of algebra in any precise or meaningful sense. As we have already mentioned, Wallis nowhere speaks of his *Treatise of Algebra* as a history, although others, like Newton, chose to do so.[34] The use of 'historical' in the title is to be understood principally as factual documentation. Correspondingly, one of the most important tasks which Wallis sets out to achieve in the *Treatise of Algebra* is to document the work of other English authors, principally Harriot and Oughtred, who for various reasons such as untimely death or reluctance to publish, had not received the recognition which he felt was due to them.

There is an unmistakable pattern both in Wallis's own work and in the declared interests of the Royal Society. Thus, the Royal Society commissioned Wallis to digest Jeremiah Horrox's (1617?–41) papers and produce a posthumous edition of his works,[35] having earlier persuaded the Danzig astronomer Johannes Hevelius to publish Horrox's observations on the transit of Venus as an appendix to his own work on the transit of

---

32   Leibniz, *Sämtliche Schriften*, Reihe 2, vol. 1, 2nd edn, p. 696.
33   Gerardus Vossius, *De quatuor artibus popularibus, de philologia et scientiis mathematicis, cui operi subjungitur, chronologia mathematicorum, libri tres* (Amsterdam: Bleau, 1650), De philologia liber, 71.
34   See Newton, *Correspondence*, vol. 3, pp. 285/286.
35   John Wallis, *Jeremiae Horroccii* [...] *Opera posthuma* (London: J. Martyn, 1673).

Mercury.³⁶ The concern for intellectual justice and for historical documentation which such efforts express are principally to be understood within the context of *historia literaria* and are not historical in the modern sense of the word. Seen in this context, Wallis's extensive coverage of Harriot is scarcely more excessive than Leibniz's promotion of the German logician Joachim Jungius (1587–1657), whose work likewise remained largely unpublished at his death.³⁷ Admittedly, there was somewhat less justification for the long chapters which Wallis devoted to Oughtred. But then Oughtred was revered by Wallis and contemporaries for the role he had played in the birth of modern mathematics in England.

Questions of style

The *Treatise of Algebra* employs a variety of literary styles. Alongside long and detailed expositions of Harriot and Oughtred there is valuable narrative on ancient Greek mathematicians and on the introduction of Hindu–Arabic numerals. When writing on classical authors Wallis draws heavily on Vossius. However, the passages on other traditions – which are regrettably short – are largely the result of his own investigations and it is here that Wallis's historiographical skills are most in evidence.³⁸ In general, Wallis presents historical development as factual, incremental growth. This form of presentation is similar to that of the Dutch humanist scholar, who appended a chronology of mathematics to his work on the nature and constitution

---

36   Johannes Hevelius, *Mercurius in sole visus Gedani, anno Christiano MDCLXI* (Danzig: Reiniger, 1662).
37   Philip Beeley, 'Experiment, Induktion, Hypothese. Leibniz' Auseinandersetzung mit dem wissenschaftlichen Nachlaß des Joachim Jungius um 1678', *Studia Leibnitiana* (forthcoming).
38   Jacqueline Stedall, *A Discourse concerning Algebra: English algebra to 1685* (Oxford: Oxford University Press, 2002), 217.

of the mathematical sciences.[39] In the chapter on the antiquity of decimal fractions, Wallis notes, for example, that the use of sexagesimal parts began to fall into disuse due to Arabic influence and describes having found in a text of the astronomer Arzachel (1029–87) the division of the radius or semi-diameter into three hundred parts. Although these were not decimals, being so many in number they needed less subdivision than sexagesimals and thus prepared the way for future development. After him, we are told, Regiomontanus (Johannes Müller, 1436–76) first divided the radius into 60,000,000 and later 'upon further consideration' into 10,000,000 parts and 'such Division of the Radius hath been since followed in all Tables of Sines; Tangents, and Secants'.[40] Arzachel is thus situated in a tradition which finds its conclusion in the work of Briggs (1561–1631), Gellibrand (1597–1637), and Oughtred.

Wallis also employs anecdotes as a means of enlivening his presentation. In the course of his demonstration of a proposition from Archimedes' *De spiralibus* he points out that he had first produced this at the request of Sir Charles Scarborough, whom he describes as a 'Dr of Physick, one of his Majesties Physicians in ordinary' and as 'a Person eminently skilled in these affairs'.[41] Moroever, he informs his readers that he transmitted this demonstration to Scarborough 'in a Letter of Novemb. 21. 1671; together with a like Demonstration of another perplexed lemma out of Archimedes'.[42] Equally anecdotal is his description in the course of his exposition of Oughtred of how he independently discovered Cardano's generalized solution to cubic equations through the study of Descartes' *Géométrie* (1637).[43] Such anecdotal remarks are valuable within the context of *historia literaria*, and can only be seen as historical in that restricted sense.

39 Vossius, *De quatuor artibus popularibus, De universae mathesios natura et constitutione liber cui subjungitur chronologia mathematicorum.*
40 Wallis, *A Treatise of Algebra*, 31.
41 Ibid., 301.
42 Ibid.
43 Ibid., 121, 175–7; cf. Rigaud, *Correspondence of Scientific Men*, vol. 2, p. 573.

Questions of plagiarism

Where Wallis does attempt to establish – or rather claim – intellectual dependencies, he becomes exceedingly weak. Superficial similarities lead him on occasion to assert not so much the borrowing of ideas as outright stealing. Time and again Descartes is presented as having taken substantial parts of the third book of his *Géométrie* directly but silently from the *Artis analyticae praxis*. Thus after outlining the principles of Harriot's treatment of cubic equations he notes:

> Which is the Mystery that before Harriot, was not (that I know of,) discovered by any. But he is since followed by Des Cartes, (but without naming him,) as well in this, as in many other things[.][44]

Nor is this approach restricted to Harriot or to the *Treatise of Algebra*. When Wallis caught sight of a curve which appears to resemble a cycloid in a Cusanus (1401–64) manuscript (or rather a later transcription), given to him by his nephew Stephen Bate and which he in turn gave to the Bodleian Library,[45] he persuaded himself without consulting the relevant text in *De complementis mathematicis* that Nicholas of Cusa had first generated the curve. So strong was his conviction that he published a paper on his discovery including an image of the manuscript in the May 1697 issue of *Philosophical Transactions*.[46]

The close of the *Treatise of Algebra* brings Wallis to contemporary analysis and allows him to situate his own work in relation to that of James Gregory (1638–75), Mercator and Newton. After outlining the classical method of exhaustion, he discusses indivisible techniques employed by

---

44   Wallis, *A Treatise of Algebra*, 135; see Rigaud, *Correspondence of Scientific Men*, vol. 2, p. 573; Beeley and Scriba, 'Wallis, Leibniz und der Fall von Harriot und Descartes'.
45   Oxford, Bodleian Library, MS Savile 55.
46   John Wallis, 'An extract of a letter from Dr Wallis, of May, 1697, concerning the Cycloeid known to Cardinal Cusanus about the year 1450; and to Carolus Bovillus about the year 1500', *Philosophical Transactions* 229 (June 1697), 561–6.

Cavalieri in his geometry of the continua and his own arithmetization of Cavalieri in *Arithmetica infinitorum*, before turning to Newton's work on tangents and series. Remarkably, he also quotes the Leibniz series, which the German philosopher and mathematician had published in the *Acta eruditorum*,[47] although he adds the remark that much contained in 'De vera proportione circuli' was within the compass of Mr Newton's general rules.[48] Nonetheless there was no suggestion of dispute in the English edition of the *Treatise of Algebra*; Wallis would later say that at the time he had only known little of what Leibniz had published in mathematics.[49] As mentioned already, the passages on Newton and Leibniz are the most heavily revised in the Latin edition of 1693, where Wallis draws substantially on Newton's letters to Oldenburg of 13 June and 24 October 1676, the so-called *epistola prior* and *epistola posterior*.[50]

Wallis and priority

By the time the Latin *Algebra* appeared, questions of chronology had become decisively important in arguing claims to priority in the discovery of the calculus. Such an approach to history was not new. Already eight years earlier in the English edition the Savilian professor had sought to establish Harriot's claim to priority over Descartes in similar fashion. Wallis's attempt to demonstrate priority and even dependency by means of the presentation of facts, much as a lawyer would present his case in court, was already established in the field of mathematics when he wrote the first draft of the *Treatise of Algebra* some time in the early 1670s. Moreover, it was

---

47   Gottfried Wilhelm Leibniz, 'De vera proportione circuli ad quadratum circumscriptum in numeris rationalibus expressa', *Acta eruditorum* (February 1682), 41–6.
48   Wallis, *A Treatise of Algebra*, 346.
49   Wallis, *Opera mathematica*, vol. 1, sig. a4^r.
50   Leibniz, *Sämtliche Schriften*, Reihe 3, vol. 1, pp. 533–54; Reihe 3, vol. 2, pp. 83–116.

an approach to history with which Wallis was exceptionally well acquainted, since he saw himself with some justification as having had his own claims to discovery in the past denied by means of factual misrepresentation.

At the end of 1658, almost immediately following the conclusion of his exchanges with Pierre de Fermat and others over problems in number theory, a series of prize questions on the properties of the cycloid, anonymously posed to the learned public by Blaise Pascal, arrived for Wallis in Oxford.[51] In contrast to the largely open nature of the debate arising from Fermat's challenges, the proceedings of Pascal's prize questions were cloaked in silence. Although the Savilian professor, keen to show that he was equal to the task, submitted a workable solution in good faith, he received no response from Paris. To make matters worse, Wallis's name was not so much as mentioned in the *Histoire de la roulette* (1658), Pascal's subsequently-published account of the competition. Of the two English contributions only Wren's determination of the arc length of the cycloid was reported, but incompletely. Having been slighted in this way, and convinced of the merit of his solutions, Wallis felt the need to defend his name in the Republic of Letters. In the first part of his *Tractatus duo* (1659), he published not only the corrected version of his own solutions to Pascal's prize questions, but also a short account of the way in which from his point of view the competition had been conducted.[52]

Not surprisingly, Wallis's account differs from that of the French mathematician and philosopher. These differences also extend to the history of the curve itself. Whereas Pascal ascribes priority in the discovery of the cycloid and its quadrature to his fellow-countrymen Mersenne (1588–1648) and Roberval (1602–75) respectively, Wallis accepts the Italian interpretation, according to which it was Galileo (1564–1642) who first discovered the curve and Torricelli (1608–47) who first performed its quadrature.

---

51   Joella G. Yoder, *Unrolling Time: Christiaan Huygens and the mathematization of nature* (Cambridge: Cambridge University Press, 1988), 77–82.

52   Philip Beeley and Christoph J. Scriba, 'Disputed Glory. John Wallis and some questions of precedence in seventeenth-century mathematics', in Hartmut Hecht *et al.*, eds, *Kosmos und Zahl. Beiträge zur Mathematik- und Astronomiegeschichte, zu Alexander von Humboldt und Leibniz* (Stuttgart: Steiner Verlag, 2008), 275–99, at 287–9.

Four years later Carlo Dati would marshall all the material to support this interpretation.[53] On the basis of available evidence, in particular Fermat's letter to Mersenne of February 1638, it is clear that Roberval already by this time was aware that the area of the cycloid is thrice that of the circle by which it is constructed. However, it was Torricelli who first published this result, namely in his *De sphaera et solidis sphaeralibus libri duo* (1644). It is this distinction between discovered knowledge and published knowledge on which Wallis focuses in the account he presents in *Tractatus duo*, where he also explicitly recognizes the possibility of parallel discoveries – something he later appears to rule out in respect of Harriot and Descartes. But concerning the discovery of the cycloid, where no English interests were involved, he turns the question of priority into one of openness in dealing with new discoveries. From Wallis's point of view neither Roberval nor Mersenne have grounds for complaint, 'since they kept their discoveries under lock and key and did not make them available to the public.' Consequently, he finds it 'absolutely unjust that they do not tolerate that others rediscover what they have hidden'.[54] In this way, Wallis rejects the accusation of plagiarism leveled against Torricelli and is later supported in this by Leibniz.

History as *Historia literaria*

Wallis's efforts to document developments, to record achievements, and to do justice to contemporaries whose work had in his view been unfairly neglected, however misguided some of his conclusions were, fits squarely

---

53   Carlo Dati, *Lettera a Filalethi di Timauro Antiate della vera Storia della Cicloide e della famosissima esperienza dell' Argento vivo* (Florence: Insegna della Stella, 1663).
54   John Wallis, *Tractatus duo. Prior, de cycloide et corporibus inde genitis. Posterior, epistolaris; in qua agitur, de cissoide* (Oxford: Lichfields, 1659), 77–8; Wallis, *Opera mathematica*, vol. 1, p. 543.

with the ethos and understanding of *historia literaria* within the Republic of Letters. Nor were he and Dati alone in their conviction that Pascal's *Histoire de la roulette* could not be allowed to stand without a complete documentation of the facts refuting it – something that Wallis had not done at all and Dati only in Italian and rather one-sidedly. Not least for these reasons Johann Gröning was persuaded to publish a full account of the *Historia cycloedis, contra Pascalium, mathematicum Gallum* as part six of his *Bibliotheca universalis* (1701).[55] Gröning, a Wismar advocate, produced this history in close consultation with Leibniz who checked his facts and offered his opinion on their presentation.[56] As the title indicates, Gröning agreed with the position adopted by Wallis and Dati, but he saw the strength of his own account precisely in its objectivity: the distance he was able to adopt to his subject-matter. In effect, Gröning's reflections on his approach express the view that history is best presented when personal considerations are left aside, as Wallis's *Treatise of Algebra* manifestly demonstrates both positively and negatively. Remarkably, Gröning also takes the established style of presentation of chronological histories as his model:

> In describing the history of the cycloid it appears not disagreeable to pursue the solemn method of recounting the histories of kings or countries, and so to begin by talking briefly about the birth, infancy, adolescence, and youth of the curve.[57]

Leibniz himself employed a strictly chronological, developmental style when, in response to Collins's collection of Newton letters, published by the Royal Society under the title *Commercium epistolicum* in 1712, he produced his own account of the development of differential calculus. In the *Historia et origo calculi differentialis*, written in autumn 1714, but unpublished at his death, Leibniz presents in many ways a different kind of historical account of the growth of mathematics, recognizing the manifold complexity of the reception and development of contemporary ideas in the

---

55   Johann Gröning, *Historia cycloidis, contra Pascalium, mathematicum Gallum* (Hamburg: Liebezeit, 1701).
56   See Leibniz, *Sämtliche Schriften*, Reihe I, vol. 13, pp. 408, 448–51.
57   Gröning, *Historia cycloidis*, 9.

discovery of the calculus, ranging from Cavalieri's method of indivisibles to Mercator's work on series. Right at the beginning of the *Historia et origo*, Leibniz presents his reasons for drawing up his historical account, which were of course not entirely disinterested:

> It is most useful that the true origins of memorable discoveries be known, especially of those, which did not become known through accident, but through the power of deliberation. For this is useful not only in order that in the history of learning each be rewarded according to his merits, and others rightfully receive their glory, but also in order that the art of discovery be enriched through an understanding of the method employed in distinguished examples.[58]

Conclusion

Although Wallis's motivation for his long expositions of Harriot and Oughtred in his *Treatise of Algebra* was undoubtedly similar to that of Leibniz's presentation of the development of the calculus in the *Historia et origo*, he lacked the historiographical tools for going beyond what the printed texts of these authors already contained. In this respect Wallis remained deeply wedded to the Aristotelian tradition in history and correspondingly failed to produce any convincing examples of intellectual dependency. When Newton calls his treatise a 'history of algebra' this reflects above all the enduring power of that tradition. As a historian of mathematics understood in the sense of causal developments Wallis was decidedly unsuccessful, while his efforts to provide a lasting memorial of the scientific achievements of earlier authors and those of his own times in the sense of *historia literaria* were outstanding.

So, was the choice of a copy of the *Algebra* for the background to his portrait such a good choice after all? There are reasons for doubt. Perhaps

---

58   Gottfried Wilhelm Leibniz, ed. C.I. Gerhardt, *Leibnizens mathematische Schriften* (7 vols, Berlin and Halle: A. Asher and H.W. Schmidt, 1849–63), vol. 5, p. 392.

it was propitious that opened out alongside the copy was a plate depicting the relation between the ellipse and the hyperbola, taken from the *Algebra* admittedly, but originating from one of Wallis's truly great works, the *De sectionibus conicis*, published a good thirty years before the English edition of his *Treatise of Algebra*.

Acknowledgements

The author should like to express two special debts of gratitude. No-one working today on Wallis's *Treatise of Algebra* can fail to have benefitted from the wealth of scholarship contained in Jackie Stedall's publications on the topic, and this author is no exception. He has also benefitted immensely from discussions with Per Landgren, whose groundbreaking work on the enduring importance of the Aristotelian concept of history in the sixteenth and seventeenth centuries is set to change many time-honoured conceptions. He is also very grateful to Benjamin Wardhaugh for providing the opportunity to present an earlier version of this chapter at the workshop on the history of history of mathematics organized by him at All Souls College, and to the participants in that workshop for their helpful comments.

BENJAMIN WARDHAUGH
ALL SOULS COLLEGE, OXFORD

'It must have commenced with mankind': some ancient histories of arithmetic in eighteenth-century Britain

> Very little is known of the origin and invention of arithmetic. In fact it must have commenced with mankind, or as soon as they began to hold any sort of commerce together; and must have undergone continual improvements, as occasion was given by the extension of commerce, and by the discovery and cultivation of other sciences. It is therefore very probable that the art has been greatly indebted to the Phœnicians or Tyrians; and indeed Proclus, in his commentary on the first book of Euclid, says, that the Phœnicians, by reason of their traffic and commerce, were accounted the first inventors of Arithmetic.[1]

This description of the early history of arithmetic, from Charles Hutton's 1795–6 *Mathematical and Philosophical Dictionary*, presents a series of distinctive or surprising features and small puzzles, whose elucidation can help to understand how mathematical history was thought about in eighteenth-century Britain. Hutton presents arithmetic as profoundly linked to practical considerations, as against the implied alternative of mathematics as fundamentally an abstract study; he yokes it to 'commerce' in particular, rather than to any of the other uses to which arithmetic was in fact put in the early modern world. His willingness to speculate about the origin of the subject in primitive societies and his confidence in writing, albeit very briefly, about its early history, sit uneasily with the paucity of sources on which his account can have been based. Finally, it is not obvious what a passage like this one is for, or who was intended to read it and profit by

---

1   Charles Hutton, *A Mathematical and Philosophical Dictionary* (2 vols, London: J. Johnson and G.G. and J. Robinson, 1795–6), vol. 1, 142–3.

it. This article will tease out some of the implications of this passage, and pursue, very briefly, the thought that the questions of how arithmetic was conceived, historicised and legitimised are useful windows onto early modern ideas about mathematics and its history.

Histories

The questions about this passage as history are perhaps the easiest to answer: Where did Hutton get this material from? What were his sources? A number of other works on arithmetic from eighteenth-century Britain contain passages – some of them rather longer and more elaborate – on the same subject; some use almost the same words. There was, indeed, a longer history of speculation on the early and medieval history of mathematics, and Hutton may have felt that this was a subject for which a degree of speculation was warranted by precedent.

A well-known example concerns the appearance in Medieval Europe of the Hindu-Arabic numerals, the use of which went by the name of 'algorism' in English until the early modern period. Some writers felt the need to attach a historical account of some kind to the numerals, and although the name of Al-khowarizmi was known to some, the matter was sometimes handled with more creativity; the name 'Algus' was given to the imaginary mathematical sage responsible for the numerals. The thirteenth-century *Roman de la Rose* placed him beside Euclid and Ptolemy, and Chaucer's *Book of the Duchess* gave the name of this 'noble countour' in the form 'Argus'.[2] By the eighteenth century Algus/Argus seems to have vanished from accounts of mathematical invention and discovery. But a similar pat-

---

2   *Roman de la Rose* 16,373; see C.S. Lewis, *The Discarded Image: an introduction to medieval and Renaissance literature* (new edition, Cambridge: Cambridge University Press, 1994), 196–7.

tern was repeated; there was now a need to account for the appearance on the scene of algebra, and 'Geber' was used to fill the role:

> ALGEBRA [*Algebre*, F. of *Algebra*, L. from *Al* excellent, and *Giabr*, *Arab*. the Name of its supposed Inventer][3]

This specific piece of information was added to Nathan Bailey's *Etymological Dictionary* (first published in 1721) for the 1728 edition, and remained there in subsequent editions up to the end of the century. Samuel Johnson was more cautious:

> A'LGEBRA. *n.s.* [an Arabick word of uncertain etymology; derived, by some, from *Geber* the philosopher; by some, from *gefr*, parchment; by others, from *algebista*, a bone-setter; by *Menage*, from *algiatarat*, the restitution of things broken.][4]

The supposed etymology from 'Geber' was complicated by the fact that at least two real individuals were known by that name to early modern scholarship: an eighth-century alchemist and a twelfth-century astronomer. But its longevity – and its appearance in Johnson – testifies to what seems to have been quite a strongly-felt need to find a plausible if not a verifiable ancestry for the (relatively) new subject of algebra.

If speculation and invention were not newcomers to mathematical history in the eighteenth century, borrowing from one's predecessors was surely a more usual technique. A very important text for the history of arithmetic in English was the 'short history' in Alexander Malcom's *New system of arithmetick* (1730).

> That Arithmetick was very early in the World, no body can doubt, because the Idea of Number arises from all things about us. In the beginning, while the Way of Living was simple, and things were in a manner common, the Knowledge of Numbers would make a small Progress: But when *Property* and *Commerce* began to be established,

---

3   Nathan Bailey, *An universal etymological English dictionary* (London: J. Darby and others, 1728), s.v. 'Algebra'.
4   Samuel Johnson, *A dictionary of the English language* (second edition, 2 vols, London: J. and P. Knapton and others, 1755–56), s.v. 'Algebra'.

Men would soon find the Necessity of enquiring into the Nature of Numbers, and contriving an *Art* of *Numbering*; without which no Business can be carried on.[5]

Like Hutton, Malcolm (1685–1763, 'Teacher of the Mathematicks at *Aberdeen*') was willing to speculate on the very earliest rise of arithmetic in human societies, and like Hutton he seems to have found it difficult to conceive of a society worthy of the name that lacked it altogether. And he shared with Hutton the belief that 'commerce' was the key driver of the first rise of arithmetic. The Phœnicians make an appearance here:

> But where, and by whom, Arithmetick received its first Form of an *Art* or *Science*, we know little about it. If the *Phœnicians* were, as it is conjectured, the first Merchants after the Flood [...] then it is probable, the *Art* began among them; by whom *Trade* and *Arithmetick* were carried into *Egypt*.

Malcolm, like many of those who followed him, went on to give a brief history of notation for numbers, dwelling on the Greek and Roman systems before describing 'The most perfect Method of *Notation*, which we now use' as the invention of 'the *Indians*', with 'the *Arabs*' relegated (all too predictably) to the status of mere transmitters. (There is no Algus here.) The story of arithmetic itself from its entry into Greece up to the early modern period traces a line through Euclid (*Elements* 7, 8 and 9), Boethius, Jordanus and Sacrobosco, through to the big names of the sixteenth and seventeenth centuries. It is a history of (extant) texts, and Malcolm attempted to create a plausible continuity for the subject between the antique and the modern periods, essentially by placing texts and authors in an orderly sequence.

(In passing, Malcolm seems to have had a penchant for this kind of historicization. In his *Treatise of Musick* he attempted in much the same way to present a version of the science of music which would be both relevant to eighteenth-century Britain but also, as far as possible, in detailed continuity and harmony with the ancient and medieval science of harmonics

---

5 Alexander Malcolm, *A new system of arithmetick, theorical and practical* (London: J. Osborn and others, 1730), xv.

as he understood it.⁶ Once again there was a premium on constructing a continuous sequence of expert authors and texts, although for the case of music the manifest changes in musical practice between the age of Euclid and that of Handel made the task very much harder.)

Many are the later writers in whose discussions of these subjects there are clear echoes of Malcolm's distinctive turns of phrase, and, as far as I have been able to determine, nearly every account of the early history of arithmetic written later in eighteenth-century Britain was at least in part a paraphrase, a précis or an expansion, sometimes at second or third hand, of what Malcolm had to say. But later authors added their own material and their own interpretations and emphases. Some cited earlier sources not explicitly mentioned by Malcolm, as we have seen in the case of Hutton's citation of Proclus; most pruned out even those references to earlier authorities Malcolm himself had supplied. Chief among those earlier authorities were the writers on mathematical history of the middle and late seventeenth century: Gerardus Vossius, Athanasius Kircher and John Wallis.

Malcolm's 'short history' occupies all of four quarto pages, but compared with those predecessors it gives an unusually extensive discussion specifically of the early history of arithmetic, and contains information which they do not. It seems that Malcolm must also have used, directly or indirectly, earlier sources which he did not name.

The possibilities are numerous, and it does not seem possible to say which sources Malcolm had consulted. Several ancient writers gave accounts, some very brief but some a little more extensive, of the origins of mathematical sciences which could have provided information used by Malcolm: Strabo⁷ and Herodotus, for instance, or Proclus in his (fifth-century) commentary on the first book of Euclid's *Elements*, mentioned

---

6   Alexander Malcolm, *A Treatise of Musick; Speculative, Practical, and Historical* (Edinburgh: for the author, 1721).
7   Anon., *An universal history, from the earliest account of time to the present* (7 vols, Dublin: George Faulkner, 1744–1745), vol. 1, p. 404.

by Hutton, where the Phoenicians and their invention of mathematics for the purpose of trade were relatively prominent.[8] He could equally have used the syntheses of relevant ancient topoi made by the sixth- and seventh-century encyclopedists Cassiodorus and Isidore of Seville. It is probably more likely that he used ancient sources than that he consulted any of the fifteenth- or sixteenth-century accounts of the ancient history of mathematics,[9] since he omitted elements which were common if not ubiquitous in them, such as the transmission of (all) learning through the 'four monarchies' – the Assyrian, Persian, Macedonian, and Roman empires – or the incident in which the sciences were saved from destruction during the Biblical Flood by being inscribed on two pillars by the children of Seth.[10] This last, taken from Josephus, produced a history which had the advantage of saving both the Biblical account of early human history and the information reported by Plato in the *Timaeus* and *Critias* about the early transmission of knowledge,[11] tracing knowledge, and the origins of arts and sciences, as far as possible back to the Biblical patriarchs, or even to God *via* Adam, but giving roles in transmission and elaboration to other ancient near eastern cultures. It was exploded in 1722 by the natural philosopher and theologian William Whiston (Newton's successor in the Lucasian chair at Cambridge),[12] and found no place in the accounts of Malcolm or his successors.

Whatever his sources were, Malcolm was not using them slavishly: compared with the ancient sources he added speculations of his own; compared with the Renaissance historians of mathematics he omitted

---

8   See Robert Goulding, *Defending Hypatia: Ramus, Savile, and the Renaissance rediscovery of mathematical history* (Dordrecht: Springer, 2010), 6–8.
9   See James Steven Byrne, 'A Humanist History of Mathematics? Regiomontanus's Padua Oration in Context', *Journal for the History of Ideas* 67 (2006), 41–61; and Nicholas Popper, '"Abraham, Planter of Mathematics": histories of mathematics and astrology in early modern Europe', *Journal for the History of Ideas* 67 (2006), 87–106.
10 Flavius Josephus, *Jewish Antiquities* 1.67–71.
11 See *Timaeus* 22b ff.
12 Popper, 'Abraham, Planter of Mathematics', 105 with note 48.

material which had now been discredited. He even omitted that favourite of mathematical historians, the periodic flooding of the Nile as a stimulus for ancient mathematical invention, a titbit which several of his eighteenth-century successors could not resist reinstating.[13] He was not, as far as I can determine, following any one source in detail, but constructing a new account suited to his own purposes.

Philosophies

What were those purposes? What agendas prompted the inclusion of the ancient historical material which Malcolm and others placed in their arithmetic books?

One was the desire to baptize arithmetic by providing for it an ancient history consistent with Biblical history. At a time when it was still possible for arithmetic to have unwelcome astrological or even cabalistic connotations[14] such a passage had the potential to comfort those nervous about approaching it. Here is the mathematical pedagogue William Butler, writing near the end of the eighteenth century: for him it seems to have been particularly important to locate arithmetical expertise among the Biblical patriarchs.

---

13   See Herodotus, *Histories*, 2.109.
14   J. Peter Zetterberg, 'The Mistaking of "the Mathematicks" for Magic in Tudor and Stuart England', *The Sixteenth Century Journal* 11/1 (1980), 83–97; compare Richard Sault, 'Of mathematics in general' in *A Supplement to the Athenian Oracle* [...] (London: Andrew Bell, 1710), 95:
> We read of many Persons, who in this Study have trod so near upon the heels of Nature, and dived into things so far above the Apprehension of the Vulgar, that they have been believed to be Necromancers, Magicians, etc., and what they have done to be unlawful, and performed by Conjuration and Witchcraft, although the fault lay in the People's Ignorance, not in their Studies.

The directions [...] given to Noah, concerning the dimensions of the ark, leave us no room to doubt that he had a knowledge of numbers, and likewise of measures. When Rebecca was sent away to Isaac, Abraham's son, her relations wished that she might be the mother of thousands of millions; and if they had been totally unacquainted with the rule of multiplication, it is impossible to conceive that they could have formed such a wish.[15]

Taking the same idea still further, some writers used their mathematical histories to express the view that all knowledge, and *a fortiori* all mathematical knowledge, had its source in God: perhaps, historically, via Adam and his immediate descendants. For Benjamin Donne, mathematical essayist of the mid-century,

> It is highly reasonable to suppose, that some Method of Numbering was used by *Adam* and *Eve* in *Paradise*, for communicating their Ideas to each other, of so many, or so much, &c.[16]

For some, such uses of mathematical history tended towards a Platonizing view of mathematics itself, in which it was not an invention born of human necessity but an emanation from the godhead. Mathematics, on such a view, would be one of the privileged possessions of those enjoying a special relationship with the deity, and transmitted from them to the rest of humanity. Donne and several other authors, including Samuel Clark, writing in Temple Henry Croker's influential 1764 encyclopedia, picked up Josephus's remark that 'Abraham taught the Egyptians arithmetic during the time of his sojourning in their country',[17] and followed the general shape of Josephus's account of the origin of sciences, in which knowledge was ultimately to be traced back to the early Biblical patriarchs and considered

---

15 William Butler, *Arithmetical questions, on a new plan* (second edition, London: for the author, 1795), 1.
16 Benjamin Donne, *Mathematical essays; being essays on vulgar and decimal arithmetic* (second edition, London: W. Johnston, P. Davey and B. Law, 1758), xxviii.
17 Samuel Clark in Temple Henry Croker, *The complete dictionary of arts and sciences* (3 vols, London: for the authors, 1764), vol. 1, s.v. 'Arithmetic'. This passage appeared in substantially the same form in various reissues of Croker's *Dictionary*, and was used as a source by some later authors, including William Butler.

as transmitted outwards from the Jewish people. (The problem of reconciling this narrative with the idea of an Egyptian or a Phoenician origin for geometry or arithmetic had not been solved by Renaissance historians of mathematics, and it was not solved in the eighteenth century.)

A different view of the philosophy of mathematics led to different historical conclusions. Virtuous humanity might be conceived to have invented arithmetic to meet practical needs:

> It is very probable, however, that it was nearly coæval with civil society itself; and that they both derived their origin from necessity, the universal parent of invention: for man, stimulated by want, that most powerful incentive to all his desires, and led by an instinctive principle to discover the means of gratification, would soon think of applying combined efforts to obtain them. But the laws of inherent rectitude would immediately dictate, that the rewards should be in proportion to the toils: hence the necessity for some means to determine that proportion.[18]

Here arithmetical knowledge seems to be restricted to those societies possessing a suitable work ethic (the anonymous author of this particular passage, written in 1792, had been a 'Teacher of Mathematics in the Royal Navy'; he was quite probably an Anglican): thus it became the privileged possession, in principle, of the virtuous.

Or, again, mathematical thinking might correspond to something in human nature that was not constrained by circumstances or culture. 'The natives of Peru in South America,' according to William Butler, 'who do all by the different arrangement of grains of maise, are said to excel any European both for certainty and dispatch with all his rules'.[19] Thus, in the view also of Hutton, with whom this chapter began, it was inconceivable that humanity, or at least social humanity, might lack arithmetic, and the implication was that thinking mathematically was in some sense part of what it was to be human.

---

18 *The first principles of arithmetic. By a late teacher of mathematics in the Royal Navy.* (London: C. Forster, 1792), 1.
19 Butler, *Arithmetical questions*, 2. Contrast Michel de Montaigne, ed. Alexandre Micha, *Essais* (Paris, 1969 [orig. pub. 1580]), vol. 1, p. 255 (essai XXXI, 'Des cannibales'), where the 'cannibales' of Brazil had 'nulle science de nombres'.

Thus one function of these accounts could be to present ideas about the nature of mathematics in an accessible fashion: mathematical history could be made to express the view that mathematics was godly, that it was a mark of virtue, or that it was simply a mark of one's common humanity.

Readers

Another function of these histories was apparently to engage and to enthuse the intended readers of the works – often arithmetic primers – in which they appeared. Writings aimed at teachers from this period regularly warn of the peril of making arithmetic 'a dull study' for the young,[20] and historical materials introduced into the exposition display a range of different strategies for avoiding that danger.

One strategy was to ask the reader to engage, in imagination, with the inventors of particular mathematical techniques. An interesting example – from outside the field of British arithmetic – is provided by the beginning of Joseph Fenn's *Algebra* (1775):

> To divide a Sum, for Example, £890, between three Persons, in such a Manner that the first may have £180 more than the second, and the second £115 more than the third.
>
> It is thus I imagine a Person would have argued, who, without the least Tincture of specious Arithmetic, attempted to solve this Problem.
>
> It is manifest that if one of the three Parts was known, the other two would be immediately discovered. Let us suppose, for Example, the third, which is the least, to be known; we must add £115 to it, and this Sum will be the Value of the second; to obtain afterwards the first Part we must add £180 to this second, which comes to the same as if we added £180 more [i.e. plus] £115, or £295, to the third.

---

20 'Mrs Lovechild' [Eleanor Fenn], *The Art of Teaching in Sport; Designed as a Prelude to a set of Toys, for enabling Ladies to Instill the Rudiments of Spelling, Reading, Grammar, and Arithmetic, under the Idea of Amusement* (London: John Marshall and Co., 1785), 51.

> Let therefore this third Part be what it will, we know that this Part, more itself together with £115, more itself again together with £295, should make a sum equal to £890.
> From whence it follows that the Triple of the least Part, more £115 more £295, or more £410, is equal to £890.
> But if the triple of the Part sought, more £410, be equal to £890, this Triple of the Part sought must be less than £890 by £410. Therefore it is equal to £480, therefore the least part is equal to £160. The second will consequently be £275, and the first or greatest £450.
> It is probable the first Analysts argued in this Manner when they proposed to themselves questions of this Nature.[21]

For Fenn, speculation about 'the first Analysts' was a way to explain and to motivate particular techniques, to answer a student's 'why' question. We do things this way because the first analysts would, surely, have approached them this way. Fenn used arguments of this kind to justify quite specific basic algebraic techniques, as the extract shows: the imagined reasoning of the 'first Analysts' indicates a quite particular sequence of algebraic operations in solving the problem. An additional function for this kind of passage is to legitimate a particular pedagogy: it is good to teach this solution because this is what would naturally occur to a sufficiently bright and motivated student unaided.

A related strategy was to hold up for admiration a legendary first founder or 'author' of mathematics. A familiar figure from versions of ancient history concerned with national identity, the legendary founder seems to have been an attractive but a somewhat problematic idea for some of these writers on arithmetic. 'Some attribute it to Seth, others to Noah, and the Turks to Enoch, whom they call Edris',[22] wrote Clark, perhaps somewhat ruefully (Renaissance predecessors including Polydore Virgil had lamented the multiplicity of alleged mathematical founders, and time

---

21 Joseph Fenn, *A New and Complete System of Algebra* (Dublin: Alex. McCulloh, 1775?), 1–2.
22 Clark, 'Arithmetic'.

had evidently not improved the situation).[23] Although several eighteenth-century arithmetical writers made tentative suggestions, inviting the reader, as we have seen, perhaps to identify with Abraham, Seth, or even Adam and Eve as the very first users of 'some Method of Numbering', none seems to have made a definite commitment to a specific individual as inventor of arithmetic. Most, from one perspective or another, seem to have concurred with the view that it must, indeed, have 'commenced with mankind'.

Another popular strategy for enthusing the unwilling young was to assert, explicitly or otherwise, that the study of mathematics was of real practical value. Examples abound, and we can point for instance to the ambitious monetary and commercial examples, sometimes involving extremely large numbers, which typically feature in arithmetic primers of the eighteenth century: in Cocker's *Arithmetic*, for instance, the student could calculate two men's profit on the sale of a tun of wine, divide up the gains and losses when four partners build a ship, and determine the proportions in which to mix gold of different degrees of purity so as to make a 17-carat product. The very first monetary example in the book involved adding various sums to a total of over £265.[24] The message – learn your mathematics well and you will grow rich – could hardly be plainer, and historical passages in which the Phoenicians were said to have invented arithmetic for that very purpose provided it with strong support. For Malcolm the Phoenicians were the 'first merchants', while Donne made them the 'first Navigators';[25] in either case they could be an inspiring example of what arithmetic could do for one, and they could also make the practical, useful art of arithmetic more approachable by distancing it from any more learned cousin: 'millions practice it as an art, who never dream of it as a science'.[26]

---

23 Goulding, *Defending Hypatia*, 10–11.
24 Edward Cocker, *A Treatise of Arithmetic* (new edition, Edinburgh: for E. Wilson, 1765), 126, 127, 138–9, 21.
25 Donne, *Mathematical essays*, xxix.
26 *The first principles of arithmetic*, 1.

These strategies for giving mathematics a human face[27] and thereby engaging the interest of the student point towards mathematics as the servant of human industry; but they were compatible, as we have seen, with a range of different positions concerning the nature of mathematics itself. Something they have in common, though, is their emphasis on the enduring nature of arithmetic, and its global reach, whether those were the products of circumstance or essential facts about the nature of mathematics. In each of these historical passages it is assumed that notations may change, monarchies may come and go, but arithmetic remains in its essentials the same activity; thus it is possible to identify it in ancient or primitive situations, and it is impossible for human beings – even Adam and Eve – to live socially without it. Equally, arithmetic crosses not just large spans of time but large distances, carrying precise information from one trading centre to another and facilitating international exchange in a way which, for instance, particular languages do not.

Why was this important to these writers and their readers? Perhaps because the art of arithmetic was in a distinctive situation compared with other mathematical disciplines at this time. It lacked the academic status of astronomy or geometry: there were no professors of arithmetic in eighteenth-century Britain. At an elementary level it lacked geometry's foundation in a standard ancient text: there was no equivalent of Euclid for arithmetic, in the sense of an ancient authority who could feasibly be used as a teaching text in the eighteenth-century. Despite its practical importance, arithmetic stood in danger of seeming a poor relation to some of its siblings within the quadrivium.

In some ways the situation of arithmetic was analogous to that of music, also one of the four traditional mathematical arts. Alexander Malcolm, for one, seems to have percived the two subjects as deserving a similar historicizing reinvention, as mentioned above. Where arithmetic was divided into

---

27 In a 1990 survey of delegates to a conference on the subject of history in mathematics teaching, the most popular reported reason for 'using history in my mathematics teaching' was to give 'mathematics a human face'. Steve Russ, paper presented at a conference on 'The History of the History of Mathematics', All Souls College, Oxford, December 2010.

a learned theoretical 'science' and a practical 'art', music was divided, sometimes by the very same writers, into a 'science' or 'theory' and a 'practice'. Both 'sciences' had some basis in classical texts, which were newly edited and studied in the early modern period, yet each had largely lost touch with its practical cousin.

Crucially, though, while practical music was a recreation, practical arithmetic had vital importance for an increasing range of professions. As one example among many we might quote the Scottish physician, satirist, and sometime mathematics teacher John Arbuthnot, who published in 1701 on 'The Usefulness of Mathematical Learning', arguing for instance that arithmetic was vital to the nation for the sake of 'increase of Stock, improvement of Lands and Manufactures, Ballance of Trade, Publick Revenues, Coynage, Military power by Sea and Land, &c', and concluding that 'it would go near to ruine the Trade of the Nation, were the easy practice of *Arithmetick* abolished'.[28] Similar arguments were made throughout the century; even Mrs Malaprop wished her daughter to have 'a supercilious knowledge in accounts'.[29] Perhaps the particular situation of arithmetic in the eighteenth century – newly important to more and more people but lacking a convincing classical pedigree – created a situation in which writers on arithmetic perceived a need for a foundation narrative of a rather particular type: one which would not just create a sense of identity through continuity with the past, but also stress the wide practical importance, the universality and the accessibility of the subject: and, according to preference, its association with godliness, virtue, or human nature. These are the characteristics which the subject needed to have, and they are the characteristics which it was given by the materials studied in this chapter.

The particular case of histories of arithmetic in textbooks and encyclopedias illuminates the ways in which mathematical history-writing could respond to audience and situation, and contribute to creating a sense of

---

28   John Arbuthnot, *An Essay on the Usefulness of Mathematical Learning* (Oxford: at the Theater, for Anth. Peisley, 1701), 27, 28.
29   Richard Brinsley Sheridan, *The Rivals*, 1.2.

disciplinary identity and purpose for particular readers. Like several of the other chapters in this volume, it shows that developing a sense of the mathematical past could involve articulating a sense of what mathematics was and of what – and whom – it was for. And it illustrates how the materials of Renaissance mathematical historiography could be put to new and perhaps surprising uses in the very different circumstances of eighteenth-century Britain.

REBEKAH HIGGITT
NATIONAL MARITIME MUSEUM, GREENWICH

The 'epitome of intellectual sagacity': Biographical treatments of Newton as a mathematician

What is there to be gleaned from a study of how Newton has been represented in biographies 'as a mathematician'? Have contemporaries and later writers not all simply agreed that, as his twentieth-century biographer Richard S. Westfall wrote, he was 'a mathematician of the first order'?[1] Yet a closer study of biographies of Newton reveals interesting subtleties in this matter that are worthy of consideration. As the meaning of the word and role 'mathematician' has shifted considerably over three centuries, so depictions of Newton have either changed in order to present him in conformity with current ideas, or he has been judged against different criteria. In addition, the individual interests and preferences of biographers have created a range of images of Newton the mathematician, which have been given more or less prominence within their overall picture. It is worth remembering that many writers and almost all readers of biographies of Newton were not themselves mathematicians or historians of mathematics. This has often made judging and presenting a highly technical achievement either impossible or undesirable. We are, however, able to consider changing views of mathematics and the mathematician. This chapter will suggest that looking at biographies of Newton, as both one of the most famous scientific heroes and an individual with a particularly complex legacy, provides a means of gaining insight into changing ideas about sci-

---

[1] Richard S. Westfall, 'Newton and his biographer', in Samuel H. Baron and Carl Pletsch, eds, *Introspection in Biography: the Biographer's Quest for Self-Awareness* (Hillsdale, NJ and London: The Analytic Press, 1985), 178.

ence and mathematics, and the relationships between these fields and with other forms of knowledge.

A brief late twentieth-century account of Newton's mathematics by John Roche reveals some interesting judgements. On the one hand, he points to the view of B. Goldstein that 'the mathematics of the *Principia* is the mathematics of a mathematical physicist, not that of a pure mathematician'. On the other, he writes that on re-reading the *Principia* 'I received a strong impression that Newton had done considerable violence to geometry', before concluding that it nevertheless 'works perfectly'.[2] These quotes signal two major themes in this chapter. Firstly, there is the question of what we mean by 'mathematician' and what biographers and readers of different eras might expect from that term. Secondly, there is the language and means by which Newton presented his published work, which had to be struggled and tussled with by his readers. In addition to these there are the particular contexts in which Newton's mathematics, mathematical development and publication became particularly relevant. Chief among these was the discussion surrounding the priority dispute with Gottfried Leibniz and his followers over the invention of calculus, which, as a theme touching on Newton's achievement, personality and morality, was particularly germane to biography.[3]

Voltaire said much when he wrote that Newton 'has very few readers, because it required great knowledge and sense to understand him. Everybody however talks about him'.[4] Newton became famous both because and in spite of the impenetrability of his mathematical work, and remained famous without detailed understanding among the wider public. Newton as a mathematician has not been much considered within works that have looked at his posthumous reputation in the eighteenth and

---

2   John Roche, 'Newton's *Principia*', in J. Fauvel et. al., eds, *Let Newton Be! A New Perspective on His Life and Works* (Oxford: Oxford University Press, 1988), 50.
3   The following two chapters, by Niccolò Guicciardini and Adrian Rice, both reflect on what the former calls the 'morbid attention' to the calculus dispute.
4   Quoted in Fauvel et. al., 'Introduction', in *Let Newton Be!*, 3.

nineteenth centuries.⁵ This reflects the fact that popular Newtonianism was largely non-mathematical in character and that the great areas of debate focused on other aspects of Newton's life and work. However, alongside these more popular accounts, there have been works that include much more technical analysis of Newton's work, aimed at expert or, later, academic audiences and which were, by the later twentieth century, making use of the publication of Newton's mathematical manuscripts.⁶ Recently, Niccolò Gucciardini has examined the 'image of Newton as a mathematician', pointing to some of the contexts in the disciplinary development of mathematics as key to changing depictions of Newton, particularly in mathematical history. I will draw on some of these points here, while shifting the discussion to consider the distinctive nature of biography.⁷

Changing views of mathematics and the mathematician

The focus on Newton 'as a mathematician', as we would view it today, is not one that would have been understood by Newton's earliest biographers and admirers. In part this was because of the focus placed on the practical and experimental side of Newton's work, particularly in optics. It was the reflecting telescope and optical experiments that had first brought him into

---

5   See Richard Yeo, 'Genius, method, and morality: images of Newton in Britain, 1760–1860', *Science in Context* 2 (1988), 257–84; Patricia Fara, *Newton: the Making of Genius* (London: Macmillan, 2002); Rebekah Higgitt, *Recreating Newton: Newtonian Biography and the Making of Nineteenth-Century History of Science* (London: Pickering & Chatto, 2007).

6   Derek Thomas Whiteside, ed., *Mathematical Papers of Isaac Newton* (8 vols, Cambridge: Cambridge University Press: 1967–1981); Richard S. Westfall, *Never at Rest: a Biography of Isaac Newton* (Cambridge: Cambridge University Press, 1980).

7   Niccolò Gucciardini, '"Gigantic implements of war": the images of Newton as a mathematician', in Eleanor Robson and Jacqueline Stedall, eds, *Oxford Handbook of the History of Mathematics* (Oxford: Oxford University Press, 2008), 707–35.

the metropolitan scientific world and, in the wider public sphere, *Opticks* was a more famous and admired work than *Principia*. Even in Cambridge, before the late eighteenth century, teaching focused more on the qualitative and natural theological parts of Newton's work than on the technical content of his mathematics.[8] Newton was celebrated in poetry, eulogies and memoirs as 'the epitome of intellectual sagacity' and his work, despite its range, was presented as the model of experimental natural philosophy. While mathematics was obviously a part of his success, his work was not admired, as that of a modern mathematician might be, for its abstract 'beauty' or its revelation of realms beyond the imaginable. Mathematics in the eighteenth century was understood as focusing on real-world problems, producing useful results and working in successful alliance with experience and experiment.

Although the terms 'pure' and 'mixed' were used in relation to mathematics in the eighteenth century, both branches were understood as dealing with measurements and quantities. As John Heard has indicated in his study of the arrival of 'the pure mathematician' as a viable category, the distinction between the two was little regarded, a fact that 'indicates that the idea of mathematics divorced from the uses of society had no currency'. Newton's success as a mathematician in the eighteenth century, in Britain especially, was best understood as having produced successful outcomes. Throughout the eighteenth and well into the nineteenth century, the terms

> 'mathematics', 'pure mathematics' and 'mixed mathematics' were still usually construed as plurals, reflecting the idea of a collection of techniques for handling problems concerning number and extension.[9]

Thus, while Newton's results were spectacular, if mathematics is understood as a 'collection of techniques' or useful tools that might be wielded in different contexts by all competent mathematicians, Newton's mathematical

---

8   Andrew Warwick, *Masters of Theory: Cambridge and the Rise of Mathematical Physics* (Chicago and London: University of Chicago Press, 2003), 57.
9   John Heard, 'The Evolution of the Pure Mathematician in England, 1850–1920', unpublished PhD dissertation (University of London, 2004), 43, 51.

work might be viewed as partially a failure. His were tools that he initially chose not to divulge, and which were never expressed by him in a way that made them easily available to others. At a period when the 'mathematician' might be a surveyor, instrument maker or mechanic, there was no ready-made box in which to place someone like Newton. It was only with the development of more abstract mathematics, first on the Continent and subsequently in Britain, that the term began to be understood as denoting a specialist practitioner of an essentially academic discipline, rather than someone who was skilled in using mathematics as a means of solving practical problems. It was not until the second half of the nineteenth century that the idea of pursuing pure mathematics for its own sake became an intelligible notion, alongside which developed the idea of mathematics as reaching beyond the material.[10]

Naturally, these developments had an impact on the way that Newton was presented, and how his mathematics were researched and judged. Newton's work could be open to one kind of criticism from those, of the eighteenth century, who saw mathematics as the creation and use of clearly described tools, and quite another from those, of the nineteenth, who were increasingly interested in the power of notation to take mathematical analysis into more and more abstract realms. Throughout, among specialists, was the knowledge that Newton had chosen an older and more cumbersome idiom in which to present his published work, a fact which made Newton's mathematical legacy much more mixed than it might otherwise have been. In more popular presentations, the development of the ideal type of the pure mathematician created different assumptions about how and why Newton focused on mathematics. This figure owed much to existing notions of scientific genius, the tropes of which included mathematical or philosophical reverie, the solitary thinker, and the mind 'voyaging on seas of thought alone', to reach barely comprehensible truths.[11] Such ideas contributed to making the Newton of the *Principia* more important than the Newton of the empirical or the speculative *Opticks*, although they

---

10   Ibid., 53.
11   The quotation is from the 1850 version of William Wordsworth's 'The Prelude'.

raised questions about whether natural ability, inspiration and genius, or dedicated and methodical hard work were the more certain or more admirable paths to scientific advance.[12]

A 'natural' mathematician

Despite the subsequent interest in his optical work, Newton's early reputation was, of course, due to his mathematical abilities, which had brought him to the attention of Isaac Barrow at Cambridge and, later, caused Edmond Halley to seek his views on orbits and gravitation. Admiration for Newton as a mathematician was, therefore, a constant theme. While the term could have humble and workman-like associations in the eighteenth century, the plaudits for Newton's abilities are numerous and unequivocal. Bernard le Bovier de Fontenelle, who, as author of the *éloges* produced by the French Académie, was Newton's first biographer, wrote of his 'sublime Notions in Geometry'.[13] Jean Baptiste Biot, another Frenchman who made a significant impact on the history of Newtonian biography, placed him 'in the first rank of discoverers in every branch of pure and applied mathematics'.[14] For William Whewell, author of *History of the Inductive Sciences*, he was 'the greatest mathematician and philosopher in Europe'.[15] Augustus De Morgan, a significant figure in nineteenth-century mathematics and history of mathematics, saw Newton as 'a combination of the greatest mathema-

---

12   See especially Yeo, 'Genius, method and morality'.
13   Bernard le Bovier de Fontenelle, *The Life of Sir Isaac Newton. With an Account of his Writings* (London: James Woodman and David Lyon, 1728), 9.
14   Jean-Baptiste Biot, trans. Howard Elphinstone, *Life of Newton* (London: s.n., 1829), 26.
15   William Whewell, *Newton and Flamsteed: Remarks on an article in number CIX of the Quarterly Review* (London: J. & J.J. Deighton, 1836), 14.

tician with the greatest thinker upon experimental truths'.[16] Despite this agreement, however, there was difference of opinion over how Newton reached his mastery, reflecting then-current debates over the nature of scientific genius, ideas about the role of mathematics, and the desire to reveal or hide Newton's debt to others.

In many biographies there was a tendency to underscore Newton's remarkable abilities by suggesting that they were largely innate or self-taught, and that he could achieve great feats with remarkable ease. Much of this view was based on what was said by Newton himself and those closest to him, for the standard biographical account of the eighteenth century was developed from Fontenelle's biography, much of which was based on a 'Memoir' by John Conduitt, the husband of Newton's niece.[17] A further source was Newton's own account of the development of his mathematical ideas, produced in response to the priority dispute with Leibniz and published in *Commercium epistolicum* (1712). While other writers and mathematicians are named, the overriding impression is of natural talent, or genius, that owed little to others. Thus most early biographies recount that Newton's natural inclination as a youth was toward books, and that it was so marked that he was removed from work on the Woolsthorpe estate and sent back to school and then university 'to follow the Bent of his own Genius'.[18]

The arrival in Cambridge was followed not by an account of his interaction with teachers, but of his personal reading, revealing the stereotypical mistake of the self-confident and unmistakably brilliant youth:

---

16 Augustus De Morgan, 'Newton', in Charles Knight, ed., *The Cabinet Portrait Gallery of British Worthies*, vol. 11 (London: Charles Knight & Co, 1846), 106.
17 Fontenelle, *The Life of Sir Isaac Newton*; John Conduitt, 'Memoir', in Rob Iliffe, ed., *Eighteenth-century Biography of Isaac Newton: the Unpublished Manuscripts and Early Texts* (Early Biographies of Isaac Newton, 1660–1885, vol. 1) (London: Pickering & Chatto, 2005), 98–106.
18 Fontenelle, *The Life of Sir Isaac Newton*, 4. See also the other brief eighteenth-century biographies transcribed and discussed in A. Rupert Hall, *Isaac Newton: Eighteenth Century Perspectives* (Oxford: Oxford University Press, 1999).

> In learning the Mathematicks, he had not the least Regard to *Euclid*, who appear'd to him too plain and easy to employ his Time about; he understood him almost before he had read him, and at his first View of the Subject of a Proposition, was able to Demonstrate it. He threw himself at once upon such Books as the geometry of *Des Cartes*, and the Opticks of *Kepler*.[19]

This soaring ability is further attested to in these writings by pointing to the early age at which Newton had made his greatest discoveries and the fact that Barrow, Lucasian Professor of Mathematics at Cambridge, had apparently resigned his chair in Newton's favour, evidently bowing out before the greater mathematician. The *Principia*, in its very difficulty and obscurity, could be presented by supporters as proof of Newton's uncommon abilities in geometry. It took time to be comprehended by others but, once understood, 'the approbation, which had been so slowly gained, became so universal, that nothing was to be heard from all quarters but one general shout of admiration.'[20] In order to demonstrate his continuing brilliance, Conduitt, Fontenelle and others presented the story of Newton finding a solution to a problem sent to mathematicians across Europe in his spare time after a day's work at the Mint: it was 'a considerable Undertaking, being intended a very difficult Proposition, yet to Sir *Isaac Newton* it was a mere Pastime.'[21]

These elements reappeared in later works, although the picture was made more subtle by the integration of further sources and in response to changing ideas about the nature and role of genius in scientific discovery. In Fontenelle's *éloge*, despite the tone of grandeur regarding Newton's abilities, we are still given the sense of his work with mathematics as representing a particular facility with a range of tools:

> Matters that almost escape from enquiry because of their subtlety could by him be reduced to calculation, such calculation as demands not only the knowledge possessed by good geometers but a peculiar skill also.[22]

---

19   Fontenelle, *The Life of Sir Isaac Newton*, 4.
20   Thomas Birch, quoted in Hall, *Isaac Newton*, 88.
21   Fontenelle, *The Life of Sir Isaac Newton*, 25.
22   Quoted from the translation of Fontenelle's *éloge* in Hall, *Isaac Newton*, 67.

By the nineteenth century, however, such biographical 'facts' were reinterpreted, with Biot's conception of Newton being close to the mould of Romantic genius. Rather than an older idea of an individual *having* a particular kind of genius, there was a new idea of *being* a genius, that could colour interpretation of a whole life and character. This biography was the first to introduce evidence for Newton's 'mental aberration' in the 1690s, alongside the idea that his original scientific work largely ceased after this date. Biot's focus on youthful vigour and inspiration, together with the idea that such brilliance risked burning itself out, reflects the Romantic ideal. In conformity with this image, his Newton was not just distracted by reading mathematical books in his youth, but 'entirely absorbed in meditation'.[23] Biot likewise insisted on the earliest dates possible for all of Newton's most important work, concluding that 'Newton made all these analytical discoveries before the year 1665, that is, before completing his twenty-third year'.[24]

Others, however, rejected this notion of scientific or mathematical genius, preferring to see a steadier hero, and a role for the progress of knowledge through pedagogy and the application of method. David Brewster, author of the longest biographical accounts of Newton produced in the nineteenth century, chose to downplay Newton's early abilities significantly. The accounts of Newton's reading and mathematical meditations were entirely replaced with the recently-published tales collected by William Stukeley of Newton's practical tinkering with windmills, dolls' furniture and kites.[25] Brewster suggested that this indicated that when Newton arrived at Cambridge, 'he brought with him a more slender portion of science than falls to the lot of ordinary scholars'. He painted this as a positive advantage, Newton's powers not having been exhausted prematurely, but it also meant that

---

23 Biot, *Life of Newton*, 2.
24 Ibid., 4.
25 Some of the biographical details collected by John Conduitt and William Stukeley, then among the Newtonian manuscripts owned by the Portsmouth family, were published in Edmund Turnor, *Collections for the History of the Town and Soke of Grantham* (London: W. Miller, 1806).

Cambridge was consequently the real birth-place of Newton's genius: Her teachers fostered his earliest studies; – her institutions sustained his mightiest efforts; – and within her precincts were all his discoveries made and perfected.[26]

Moreover, after he left, 'his disciples kept up the pre-eminence of their master's philosophy' and ensured Cambridge's distinguished place in Europe thereafter. Such an account gives a very different impression to the reader of the nature of mathematical work and the type of person who would be considered the greatest of mathematicians. It was entirely rejected by Biot, who believed Brewster had misrepresented Newton's character, as *the* epitome of scientific genius.[27]

Obscurity and mathematical notation

As suggested above, Newton's facility with mathematics and geometry could be presented as the reason why people found the *Principia* so difficult to understand. As Fontenelle wrote,

> If the intelligent Beings, superiour to Man, do also make a Progress in Knowledge, they fly whilst we creep, and pass over without Notice the intermediate Steps, by which we slowly advance from the Perception of one Truth to another, which has a Dependance upon it.

Yet he also noted that in the *Principia*

> the Author has been very sparing of his Words, and the Conclusions are often very hastily drawn from the Premises, and the Reader is sometimes forced to supply an intermediate chain of Consequences.[28]

---

26 David Brewster, *The Life of Sir Isaac Newton* (London: John Murray, 1831), 13.
27 See also Fara, *Newton: the Making of Genius*, 216–17.
28 Fontenelle, *The Life of Sir Isaac Newton*, 5, 8.

The 'epitome of intellectual sagacity'

This, which is notably critical for a eulogy, reflected the wider criticism and dissatisfaction with the *Principia* that Gucciardini has noted in scientific circles.[29] While incomprehensibility might mark the distance between Newton's extraordinary intellect and that of the man with merely ordinary mathematical skills, it might also be seen as a significant problem for science as a collective enterprise.

Mathematicians, struggling with Newton's omissions and obscurities, and his focus on results over process and method, certainly made negative comments, but less of this is made plain in biographies, which tended to be celebratory and aimed at audiences that included non-specialists less well equipped to understand the issues at stake. As a Frenchman, Fontenelle was in a position to highlight controversial areas in Newton's biography and science, where British authors might tread a more cautious path. There was certainly nothing in eighteenth-century English biography that laid out the kind of criticism noted in private or in mathematical commentaries. Nor, indeed, were biographers likely to follow the line of Joseph Priestley, who suggested that Newton and his followers had deliberately obscured steps in his work, and missed out the role of error, guesswork and accident, in order to place him higher on his pedestal. The truth, he suggested, would mean 'our amazement at the extent of his genius would a little subside', which would, in his view, be beneficial to science and mankind by encouraging wider participation.[30] It was not until the nineteenth century that biographers and biographical commentary on Newton began to tackle this theme, by which time the British Newtonian mathematical legacy had been largely supplanted by the introduction of Continental analytical techniques. Newton's mathematics were now seen as cumbersome, wildly out-dated and responsible for Britain's apparent backwardness in mathematical sciences compared to France.[31]

---

29 Gucciardini, 'Gigantic implements of war'.
30 Joseph Priestley, *The History and Present State of Electricity* (London: Printed for J. Dodsley, J. Johnson and T. Cadell, 1767), quoted in Higgitt, *Recreating Newton*, 7.
31 Gucciardini, 'Gigantic implements of war', 734–5.

In biography, this theme was the specialist subject of Augustus De Morgan. He was motivated by these new conceptions of what was admirable in mathematics, as well as by reasons that will be returned to in the following section and in Adrian Rice's chapter in this volume. In an 1846 biography he wrote that he was 'avowedly expressing, in one point, [his] low estimate of Newton's power', with regard to his invention and use of mathematical language:

> With few advantages as a writer or teacher, he wraps himself in an almost impenetrable veil of obscurity, so as to require a comment many times the length of the text before he is easily accessible to a moderately well-informed mathematician.

As De Morgan noted, 'the genius of Newton did not shine in the invention of mathematical language' and therefore 'the idea of the *permanent use of an organized mode of mathematical expression*' had been borrowed from Leibniz.[32] As he wrote elsewhere:

> The matter of Newton's mind and writings bears out this view. *He was not a master of expression*: in all his writings he is obscure and clumsy in his handling of tools [...] I should as soon have expected Leibnitz to have worked out gravitation as Newton to have originated any thing worthy of himself in notation.[33]

Despite his criticism here, and in other topics relating to Newton's life and character, De Morgan found his explanation, as Fontenelle did, in Newton's great abilities:

> He did not cultivate a crop for which he had no use. He who can make existing language serve his purpose never invents more: and Newton was able to think clearly and powerfully without much addition to the language he found in use. The Principia,

---

32 De Morgan, 'Newton', 95.
33 Draft letter from De Morgan to Brewster (1842), in Rebekah Higgitt, ed., *Nineteenth-Century Biography of Isaac Newton: Public Debate and Private Controversy* (*Early Biographies of Isaac Newton, 1660–1885*, vol. 2) (London: Pickering & Chatto, 2005), 176. This seems to be in response to a letter from Brewster to De Morgan of 6 August 1842 (London, Royal Astronomical Society, De Morgan MSS).

obscure as it is, was all light in Newton's mind: and he did not attempt to conquer difficulties he never knew.[34]

If it required the work of others to translate and rework the *Principia*, this did not lessen what Newton alone could achieve and, by the 1840s, when this translation was fully embedded within British teaching and research, the problems of that legacy had largely been overcome by British mathematicians.

The biographer as subject

While there are general trends over time that may be traced through the biographies of famous people, individual interests are also implicated. Some of these, naturally, relate to the time and context of writing, but there are also more personal choices. Here I will focus on some of the key nineteenth-century biographers of Newton and pick out aspects of their representation of Newton as a mathematician that are peculiar to their interests. National and disciplinary loyalties are particularly apparent, but differing views of the importance, or necessity, of including sophisticated mathematics in scientific work, of the role of notation within mathematics, the role of history of mathematics, and ideas about what does and does not count as science might all play a role. This section therefore looks in greater depth on three of the authors already mentioned: Biot, Brewster and De Morgan.[35]

---

34   De Morgan, 'Newton', 96.
35   That biographies tell us more about the biographer than the subject is, of course, something of a commonplace, and underpins much of the argument in Higgitt, *Recreating Newton* (2007). With regard to Newtonian biography, the point was made cogently in Paul Theerman, 'Unaccustomed role: the scientist as historical biographer – two nineteenth-century portrayals of Newton', *Biography* (1985), 145–62. See also

As already noted, Biot's depiction of Newton was influenced by the idea of Romantic genius being applicable to scientific as well as artistic endeavour. However, he was interested in ensuring that Newton did not stand alone, as the genius so often does, for he aimed to use this great reputation to bolster that of the French tradition of mathematical physics with which he identified himself. I have discussed this elsewhere with regard to Biot's optics, but it inevitably has important consequences for his treatment and understanding of Newton's mathematics.[36] Biot was concerned to demonstrate that Newton's real legacy had not been developed within Britain, even if British mathematicians had remained faithful to Newtonian fluxions, but on the Continent and through a lineage leading directly to Pierre-Simon Laplace and his followers. He went as far as to write of Laplace's 'genius' as that which 'has so much contributed to Newton's glory'.[37] Laplace and Biot agreed that it was the defects of Newtonian fluxions that were the chief limitation of the *Principia*, but also that the 'unavoidable defects' of this undeveloped tool had been corrected, and their power vastly extended, by Continental mathematicians over the following century. Biot was certain that, despite the geometrical presentation of the *Principia*, it was 'evident' that 'Newton attained these great results by the help of analytical methods, of which he had himself so much increased the power'.[38]

In tying Newton's achievements to the Laplacian worldview, Biot was, however, also claiming prestige for a view of mathematics that, like Newton's, was concerned with physical problems. This was in part a defensive measure against the more abstract tendency of recent mathematical work. Biot's Newton was no pure mathematician, but one who brought the rigour and certainty of mathematics into experimental physics, and one who was, therefore 'le créateur de la philosophie naturelle, l'un des plus grandes promoteurs de l'analyse mathématique, et le premier des physiciens qui ont

---

the comments by Adrian Rice, in his examination of De Morgan's motivations in rehabilitating Leibniz's reputation, and by Henrik Kragh Sørensen, in this volume.

36 Higgitt, *Recreating Newton*, 20–4.
37 Biot, *Life of Newton*, 23.
38 Ibid., 24.

The 'epitome of intellectual sagacity'   61

jamais existé' [the creator of natural philosophy, one of the chief promoters of mathematical analysis, and foremost among the physicists that have ever existed].[39] We might well question the idea of Newton, reluctant to publish and to share his methods, as a 'promoter' of mathematical analysis but, for Biot, the promise of what was yet to come was latent within Newton. He declared it

> impossible to enumerate the various discoveries in mathematical analysis, and in natural philosophy, that this calculus has given rise to; it is sufficient to remark, that there is scarcely a question of the least difficulty in pure or mixed mathematics that does not depend on it, or which could be solved without its aid. Newton made *all these analytical discoveries* before the year 1665 [...].[40]

While the 'calculus' referred to is that of Leibniz, he is here only presented as having 'again discovered, and presented to the world in a different form' the idea that first came from Newton.[41]

David Brewster's 1831 *Life of Newton* was in part written in response to Biot's biography, particularly with regard to Biot's claims that Newton had suffered a breakdown and that religion was an interest that post-dated this event. However, it was the most substantial biography yet produced and contained a great deal more than this polemic. Like Biot, Brewster's principal interest was in optics, although his approach was almost entirely empirical. A glance at the contents is revealing: of the nineteen chapters, six were devoted to optics and only one to mathematics, a topic that was more difficult for Brewster either to comprehend fully or to present to

---

39 [Jean-Baptiste Biot], 'Newton (Isaac)', in L.G. Michaud, ed., *Biographie universelle, ancienne et moderne*, vol. 31 (Paris: Michaud Frères, 1822), 169. The 1829 translation unsurprisingly misses out the final clause, since there was then no English equivalent of 'physicien'. It also chose to describe Newton as '*almost* the creator of Natural Philosophy' (my emphasis), perhaps leaving room for British reverence for Francis Bacon. Biot, *Life of Newton*, 26.
40 Biot, *Life of Newton*, 4 (my emphasis).
41 Ibid.

a broad readership.⁴² In reviews by experts Brewster was chided for this imbalance, while Biot criticised him for putting Newton's discoveries in the order that he published them rather than the order in which he made them, indicating a failure to understand the fundamental importance of mathematics to the whole of Newton's work and, therefore, the true significance of Newton's legacy.⁴³ Brewster's account was, indeed, largely a simple background and a brief description of the published mathematical papers. It is encountered by the reader only after the descriptions of Newton's optics and astronomy and there is little sense that this work underpinned or defined Newton's science. Although Brewster pointed to the triumph of the laws that Newton defined, the use of mathematics within these other fields, and the need for excellent mathematical knowledge to read the *Principia*, there is little indication of the mathematical physics that Biot saw as Newton's signature contribution.

As perhaps became a popular account, Brewster's description of the *Principia* avoids the technicalities in favour of the idea of the 'beautiful simplicity of the system which it unfolds'.⁴⁴ It is interesting, in this context, that Brewster, as well as describing how 'the mathematical principles of the Newtonian system were ably expounded in our seats of learning' in the decades after publication, takes time to mention some of the popularisers too, including Desaguliers and his lecture-demonstrations, 'accommodated to the capacities of those who were not versed in mathematical knowledge'.⁴⁵ Readers of Brewster's biographies of Newton required some mathematical knowledge to follow his account of Newton's mathematical

---

42 Brewster, *Life of Newton*; David Brewster, *Memoirs of the Life, Writings and Discoveries of Sir Isaac Newton* (Edinburgh: Thomas Constable & Co., 1855).

43 Jean-Baptiste Biot, 'Revue de *The Life of Isaac Newton*', *Journal des savants* (1836), 264–5. Note that here Biot also takes a different view on Leibniz and Newton. Whereas his biography essentially presented a case of simple simultaneous discovery, here he criticised Brewster for suggesting that the fluxions and calculus were analogous, rather than containing significant differences in method.

44 Brewster, *Life of Newton*, 169.

45 Ibid., 177, 180.

work, but his book is a useful reminder that popular ideas of Newtonianism were very largely stripped of technical mathematical content. Brewster's second, expanded biography of Newton also received criticism for the lack of focus on the role of mathematics. Brewster had benefitted from access to Newton's manuscripts in the possession of the Portsmouth family, and so Biot and De Morgan were disappointed that he had not published any mathematical manuscripts.[46] Brewster's main concern, however, was to get the detail and chronology right in order to continue his defence of Newton in the calculus controversy. Beyond making use of the biographical notes collected by Conduitt, and briefly admitting Newton's alchemical interests and religious unorthodoxy, Brewster paid scant attention to the remainder of the archive.

Brewster's interest in the legacy of other scientific figures and, especially, his desire to establish debts and priorities, meant that his biographies contain a significant amount of history of science and mathematics. However, Newton's relationship to his forebears and contemporaries was given more sophisticated, if less sustained, consideration by De Morgan. His criticism of hero-worship and reconsideration of the contribution of less well-known writers, together with his sense that Newton and his allies had deliberately obscured the real history of the calculus dispute, accounts for his criticism of Newton's ability in mathematical notation.[47] While Newton's gifts were great, he in fact

> serve[d] to illustrate what a popular reader would hardly suppose, namely, that the wonder of great discoveries consists in there being found one who can accumulate and put together many different things, no one of which is, by itself, stupendous.[48]

---

46 See especially Jean-Baptiste Biot, 'Revue de *The Life of Isaac Newton*', 156–66, 205–23, 641–58.
47 See Adrian Rice's comments in the conclusion to his chapter in this volume, where he suggests that De Morgan's criticism of Newton and rehabilitation of Leibniz were motivated by his sense, as a religious nonconformist, of being an outsider and against vested establishment interests.
48 De Morgan, 'Newton', 108.

This view opened up the number and range of individuals who might contribute effectively to science, even if it still required a Newton to make a significant aggregation of the efforts of his predecessors and contemporaries. Paying attention to the little men of mathematical history was, as Joan Richards has discussed, important for De Morgan in terms of understanding mathematics and how it had developed. It was also a moral duty, for the historical record was too often lost and rewritten in favour of great men.[49] De Morgan went much beyond Brewster's linear view of scientific progress in demonstrating that science and mathematics have long, populated histories and that even Newton had many scientific debts. For example, he noted that while the only text that Newton was said to have read before arriving at Cambridge was Sanderson's *Logic*, his research suggested 'Newton was not only conversant with *Barbara, Celarent*, &c., but even with *Fecana, Cajeti, Dafenes, Hebare, Gadaco*, &c.'[50]

It was a view De Morgan developed through extensive reading of historic mathematical texts, in which he was better versed than all but a handful of historians of mathematics among his contemporaries. Although he sadly never developed the idea, De Morgan hinted more than once that he was attracted to the notion that Newton

> had more acquaintance with the mode of thought of the schoolmen than any ordinary account of his early reading would suffice to explain. We strongly suspect that he made further incursions into the old philosophy, and brought away the idea of fluxions, which had been written on, though not in mathematical form, nor under that name.[51]

De Morgan's reverence for Newton's abilities, as a mathematician above all, was sincere. It was a reputation, he constantly asserted, that could not be tarnished by admitting debts or moral shortcomings in the man himself.

---

49 Joan L. Richards, 'Augustus De Morgan, the history of mathematics, and the foundations of algebra', *Isis* 78 (1987), 7–30.

50 Augustus De Morgan, 'Review of Sir David Brewster's *Life of Newton*', *North British Review* 23 (1855), 312.

51 De Morgan, 'Review of Brewster's *Life of Newton*', 312–13.

The 'epitome of intellectual sagacity' 65

> Newton, whose sagacity in pure mathematics has an air of divination, who has left statements of result without demonstration, so far advanced that to this day we cannot imagine how they were obtained, except by attributing to him developments of the doctrine of fluxions far, far beyond what he published, or any one of his time.[52]

There was, in addition, something instinctive and not entirely controlled about this ability: 'he hunted rather by scent than by sight'.[53] Yet De Morgan's view of how science progresses, and his aim of demystifying the persona and achievements of Newton, tended to counteract the effect of such passages.

Newtonian biography in an era of professional history of science

In the later nineteenth and twentieth centuries, Newtonian scholarship developed along a number of distinct paths. Firstly, Newton's scientific manuscripts, as defined by late nineteenth-century scholars, were separated out from the personal, theological and alchemical manuscripts and brought to Cambridge.[54] Their accessibility allowed for specialist research in the history of science and, especially, mathematics to develop, at a remove from the complexities of the personal archive. Beginning with W.W. Rouse Ball's *Essay on Newton's "Principia"* (1893), flourishing from the 1950s within the 'Newton industry' and culminating in Whiteside's edition of Newton's mathematical papers (1967–81), a thorough and technical understanding of Newton's mathematics has been developed. At the same time, there

---

52 Ibid., 318–19.
53 Ibid., 319.
54 See Rob Iliffe, 'A "connected system"? The snare of a beautiful hand and the unity of Newton's archive', in Michael Hunter, ed., *Archives of the Scientific Revolution: the Formation and Exchange of Ideas in Seventeenth-Century Europe* (Woodbridge: Boydell Press, 1998), 137–57.

was an increasing interest in the content of the remaining 'non-scientific' manuscripts. Famously, their sale in 1936 prompted John Maynard Keynes to present Newton as the 'last of the magicians' rather than 'the first and greatest of the modern age of scientists'.[55] Scholarship likewise turned to Newton's alchemy and religion, his debts to earlier philosophy and ideas about magic, and his work at the Mint.[56] As Westfall wrote in 1976, these specialist studies were at the expense of biographical coherence: 'The Newton of the textbooks has come unglued. As yet, no one has re-assembled the pieces into a coherent form.'[57]

Biographies of Newton, aimed at wider audiences, have naturally had varying relationships with this body of scholarship. While biography undoubtedly drove Newtonian scholarship in the early and mid-nineteenth century, there were always, alongside, scores of more popular texts that tended to present uncomplicated and idealised versions of Newton's life. Such works continued to appear throughout the twentieth century, although more ambitious biographies might draw on sophisticated analyses of Newton's work. Worthy of note are Louis Trenchard More's *Isaac Newton: a Biography* (1934) and Frank Manuel's *A Portrait of Isaac Newton* (1968). Both works took cognizance of the archive and contemporary scholarship, but in many ways are most remarkable for their differences. The former presents Newton as a mathematical physicist and 'personification of the scientific method', held up as a model against which the 'unrestrained

---

55 John Maynard Keynes, 'Newton, the man', in idem, ed. G. Keynes, *Essays in Biography* [...] *New Edition with Three Additional Essays* (London: Rupert Hart-Davis, 1951), 310–23, at 310.
56 Key works in creating a new vision of Newton were James E. McGuire and Piyo M. Rattansi, 'Newton and the "Pipes of Pan"', *Notes and Records of the Royal Society* 21 (1966), 108–43, and Betty Jo T. Dobbs, *The foundations of Newton's alchemy or 'The Hunting of the Greene Lyon'* (Cambridge: Cambridge University Press, 1975). A useful starting point for understanding the breadth of literature on Newton in the late twentieth century is Derek Gjertsen, *The Newton Handbook* (London and New York: Routledge & Kegan Paul, 1986). See also the Newton Project's bibliography at <http://www.newtonproject.sussex.ac.uk/prism.php?id=90>.
57 Richard S. Westfall, 'The changing world of the Newtonian industry', *Journal of the History of Ideas* 37 (1976), 175–84, at 176.

speculation' and 'pure symbolism' of contemporary figures were judged.[58] Manuel's biography, on the other hand, applied psychological techniques to the evidence regarding Newton's personality. This extended interest in the man and his foibles, as well as the range of his interests, including religion and alchemy. It also lent weight to the notion of genius, including mathematical genius, as a manifestation of disordered personality.[59]

It was Westfall's biography of 1980, however, that supplanted Brewster's 1855 volumes as the most complete and thorough account of Newton's life. He was able to take advantage of the full range of scholarship produced by professional historians of science, including the most technical, and to respond to the work of earlier biographers. Westfall was prepared, and indeed required, to 'admit' to faults in both Newton's personality and his mathematics and its presentation. With regard to the latter, he showed that Newton had sometimes adjusted his calculations to fit his theories, and took on Whiteside's description of the *Principia*'s 'logical structure' as 'slipshod' and 'its level of verbal fluency none too high, its arguments unnecessarily diffuse and repetitive and its very content on occasion markedly irrelevant to its professed theme'.[60] As with De Morgan, Westfall's reverence for Newton's achievements and abilities in science was unshaken, seen as separate from concerns about his moral life or 'non-scientific' interests. Not all contributors to the Newton Industry were so sanguine about diluting the tradition vision of Newton's achievement, as Westfall suggested in a wonderful footnote:

> Whiteside's voice went up a good octave in his indignation over Manuel's psychoanalysis of Newton's sexual repressions. I have not yet heard his reaction to Mrs. Dobbs' book on alchemy; he may have gone right off the audible range.[61]

---

58 Quotes from More's biography in Paul Theerman, 'Unaccustomed role', 157.
59 Frank Manuel, *A Portrait of Isaac Newton* (Cambridge, MA: Harvard/Belknap Press, 1968).
60 Quoted in Roche, 'Newton's *Principia*', 59, 50.
61 Westfall, 'The changing world', 181.

However, when Westfall was persuaded to reflect, for a symposium on 'Introspection in biography', on the extent to which he, as biographer, projected himself into his account he was somewhat surprised by the results.[62] He found that his views of the role of science, its relationship to religion and a moral sense of duty in relation to both, affected his interpretation and judgement of Newton more than he had anticipated. Of particular relevance here is the extent to which he found himself able to identify with certain aspects of Newton's output and not with others, which was revealing of his twentieth-century notion of what science is and what kinds of work and thought it entails. While Brewster had admitted the existence of alchemical writings among Newton's papers, he shuddered, passed judgement and attempted to excuse his hero. Westfall, however, had to take the more historicist approach of the professional historian of science although, as he candidly admitted, this was still profoundly difficult.

> It is with alchemy that Newton most eludes me. I can compose facile formulas to explain his participation in the art; I can continue to repeat them. There is, however, no sense in which I understand, in the full meaning of the word, how the author of the *Principia* became engaged so extensively in an enterprise that differed so profoundly from his other scientific work.[63]

Since the 1980s, general interest in this alchemical, religious and mystic Newton has often eclipsed Newton the mathematical physicist, as Michael White's biography, *Newton: the Last Sorcerer* (1998), and the BBC's *Newton: the Dark Heretic* (2003) would suggest. In other interpretations, the vision of Newton the mathematician has been brought into conformity with Newton the alchemist. James Gleick's 2003 biography sought to show how Newton's mathematics arose from the older alchemical worldview, rather than pointing forward to modern scientific disciplines.[64] It appears too that abstract mathematics, mathematical 'codes' and the sense of an almost

---

62   Westfall, 'Newton and his biographer'.
63   Ibid., 178.
64   James Gleick, *Isaac Newton* (London: Fourth Estate, 2003).

mystical revelation of the underlying 'truth' of nature have a new cultural cachet within popular science writing and broadcasting.[65]

Conclusions: biography and the history of mathematics

Newton's image as a mathematician has developed over time, 'reshaped according to agendas that were polarizing debates amongst practising mathematicians'.[66] As Gucciardini also shows in this volume, accounts of Newton's mathematics in mathematical texts and histories were influenced by efforts to understand the differences between the contributions of Newton and Leibniz, and in taking sides in this conflict. Likewise, during the early nineteenth-century, developments in calculus, dubbed the 'rigourization of analysis', and, subsequently, the development of synthetic geometry, encouraged first negative then more positive appraisals of Newton's legacy.[67] Gucciardini's analysis, based on histories of mathematics, would seem to be broadly confirmed by the biographies discussed here, especially at the crucial periods when the Continental and British mathematical worlds were at their most diverse, and as analysis was finally brought into Britain. It is no coincidence that a reinterpretation of Newton's life and legacy took place at about the same time, and that differing national traditions played their part.

However, the biographies also reveal other important influences on the image of Newton as mathematician. Alongside the development in content, there was an associated development in the role and understanding of the figure of the mathematician. This, as Heard has described, was

---

65   A recent example is the 2011 BBC television programme, 'The Code', presented by mathematician Marcus De Sautoy: <http://www.bbc.co.uk/programmes/b00zs6sl>. The headline focus on stories, secrets and mysteries, and the nod to Dan Brown's *The Da Vinci Code*, seems to link well with recent popular depictions of Newton.
66   Gucciardini, 'Gigantic implements of war', 731.
67   Ibid., 728.

socially determined and cannot be explained fully by the internal development of mathematical ideas, for the means by which mathematicians could viably make a living, and the expectation about what kind of person and personality the mathematician might be, also link to the external forces of institutions, funding and audience.[68] Scientific genius was a concept that, as Patricia Fara has shown, developed alongside the perception of Newton as an exemplar of such genius, but the role of mathematics – as opposed to such qualities as perception, concentration or penetration – was not initially obvious.[69] Through figures such as Newton, Einstein and Hawking, the modern idea of scientific genius has become closely associated with the application of the highest mathematics to understanding the universe. Influenced by the evolution of the pure mathematician as a type, mathematics is, in popular accounts, an abstract, beautiful and largely incomprehensible end in itself. There is much in Biot's account of youthful, unworldly and tortured genius that remains with us today.

The biographies discussed here also reveal a number of other facets that raise interesting questions for perceptions of mathematics and for the relationship between social history of science, history of mathematics and biography. On the one hand, there is a fairly bare account of Newton's mathematical work from Brewster, reminding us of the problems for non-experts in explaining technical work to popular audiences, and the fact that biography is often understood as the most accessible means of understanding history of science. Newton's reputation was always shaped and reshaped in such popular forums, where there was often little focus on his mathematical work and its relationship to his more spectacular discoveries. On the other hand, both De Morgan and Westfall, for different reasons and in different contexts, had an allegiance to a fully historicised understanding of Newton's mathematics and its relationship to his other work. De Morgan was unwavering in his admiration of Newton's achievement and abilities but, as a mathematician who believed lessons could be learned from accurate history and as a teacher of mathematics, he was interested in

---

68   Heard, 'The Evolution of the Pure Mathematician'.
69   Fara, *Newton: The Making of Genius*.

probing Newton's weaknesses and searching the historical record for other contributors. Even more so than De Morgan, Westfall, of course, had an allegiance to the history of science as a discipline. However, his admiration for the achievements of modern science, seen as the most important aspect of Newton's legacy, shaped the figure he presented in his biography.

While the history of mathematics has always had a strong biographical flavour, with biographies of mathematicians forming one of its principal outputs, it is interesting to note that there has been relatively little critical consideration of mathematical biography, compared to the amount of recent work on scientific biography.[70] If it is the case that mathematics has typically been treated as 'the one body of knowledge that is immune from those cultural influences that, in the history of science generally, are now seen as essential in providing a rounded account of historical episodes', then its role in biography and its usefulness in understanding the personality and its relationship to wider society are, perhaps, uniquely problematic.[71] Considering the history of the depiction of mathematics and mathematicians in biography may be one useful means to demonstrate that views of mathematics have a history that is dependent on factors outside its internal development. It may also encourage historians of mathematics and biographers of mathematicians to reflect on the relationship between mathematical ideas and the social, cultural and personal contexts of those who held them.

70 See, for example, Michael Shortland and Richard Yeo, eds, *Telling Lives in Science: Essays on Scientific Biography* (Cambridge: Cambridge University Press, 1996), and Thomas Söderqvist, ed., *The History and Poetics of Scientific Biography* (Aldershot: Ashgate, 2007).
71 Heard, 'The Evolution of the Pure Mathematician', 37.

NICCOLÒ GUICCIARDINI
UNIVERSITÀ DI BERGAMO

The Quarrel on the Invention of the Calculus in Jean E. Montucla and Joseph Jérôme de Lalande, *Histoire des Mathématiques* (1758/1799–1802)

The dispute between Newton and Leibniz concerning priority in the invention of the calculus is a very well researched chapter in the history of mathematics. This notorious squabble has never ceased in fact to attract a somewhat morbid attention. From the point of view of the historiography of mathematics, it is a very promising case study, not only because the amount of material concerning the dispute allows for comparative analyses of the various historical narratives woven across three centuries, but also because different historians have examined the question from different viewpoints.[1] Nationalistic issues, of course, have played a prominent role. Even more interestingly, perhaps, what is at stake here is the question of what we mean by 'inventing the calculus', and of what criteria should be

---

1 A good starting point for gathering information on Newton's biographies is Robert Iliffe and Rebekah Higgitt, eds, *Early biographies of Isaac Newton, 1660–1885* (2 vols, London: Pickering & Chatto, 2006). Rebekah Higgitt's chapter in the present volume provides an analysis of the plurality of images of Newton as a mathematician that emerge from his biographies throughout the centuries. As she observes, the priority dispute is a theme which touched not only Newton's achievement, but that was also instrumental in defining a narrative on his morality and personality. One of the most prolific and irreverent nineteenth-century Newtonian biographers was Augustus De Morgan; Adrian Rice's chapter in this volume details the agenda that led De Morgan to view the priority dispute in a way that fostered a 'sea-change' in British attitudes towards Newton the mathematician. See also Adrian Rice, 'Augustus De Morgan: historian of science', *History of science* 34 (1996), 201–40.

invoked in order to adjudicate priority in mathematical discoveries. These are all far from trivial questions.

Establishing what one means by the discovery of the calculus has always been controversial, for three reasons. First, several mathematicians other than Newton and Leibniz contributed to its birth: Johannes Kepler, Blaise Pascal, James Gregory, John Wallis, Ehrenfried Walther von Tschirnhaus and Isaac Barrow, to name just a few. Second, a plurality of versions and notations of the calculus were available at the beginning of the eighteenth century. At least two forms may be distinguished: the calculus could be based either on infinitesimal concepts or on limits. There were also several competing notations, which broadly speaking fall into two groups: the fluxional and the differential/integral notation. Third, the calculus (at least up until Euler) was never thought of as being completely independent from geometrical representation. Calculus was far from being understood as an abstract un-interpreted formalism; geometric 'equivalents' of the algorithm were deployed in different ways and for different purposes. This was the case, for instance, with exhaustion methods reminiscent of Archimedes' work, with Newton's geometry of first and last ratios, and with Huygens' geometry of infinitesimal magnitudes.

In this chapter I will discuss the historical narrative of the priority dispute as presented by Jean Étienne Montucla in his bulky *Histoire des mathématiques*. This book appeared in two tomes in 1758 and was later republished, with two additional volumes, between 1799 and 1802, thanks to Jérôme Lalande's editorial effort. My purpose is to identify the somewhat implicit assumptions behind Montucla's historical study of the priority dispute. As we shall see, Montucla's narrative of the priority dispute displays many features that are typical of the mathematical culture of some *académiciens* mathematicians and encyclopedists active in France in the latter half of the eighteenth century. But let me first introduce you to Montucla and his *Histoire*.[2]

---

2   On the life and work of Montucla, George Sarton, 'Montucla (1725–1799): His Life and Works', *Osiris* 1 (1936), 519–67, is still a useful reference. Montucla's *Histoire* is dealt with by Noah M. Swerdlow, 'Montucla's legacy: the history of the exact sciences',

Jean Étienne Montucla was born in Lyon on 5 September 1725, from a family of small traders. He attended the Jesuit College in his hometown, and then enrolled in law school in Tolouse. After completing his studies, he moved to Paris. Here he became an *habitué* of the bookseller Jombert. Booksellers provided a venue for the gathering of learned men and women of letters and science; it was at Jombert's that Montucla met some of the protagonists of the French Enlightenment: the encyclopedists Denis Diderot and Jean d'Alembert, the mathematician Jean Paul Gua de Malves (who also played a prominent role in the *Encyclopédie*), the architect Jacques François Blondel and the astronomer Jérôme Lalande. Lalande, himself a student of the College in Lyon, was to become Montucla's lifelong friend. Montucla developed an interest in the history of mathematics early in life, and his first work was an *Histoire des recherches sur la quadrature du cercle*, which Jombert published in 1754. The publication of this book, and d'Alembert's

---

Journal of the History of Ideas 54 (1993), 299–328; Jeanne Peiffer, 'France', in Joseph W. Dauben and Christoph J. Scriba, eds, *Writing the History of Mathematics: Its Historical Development* (Basel: Birkhäuser, 2002); and Pierre Crepel and Alain Coste, 'Jean-Étienne Montucla, *Histoire des mathématiques*, second edition (1799–1802)', in Ivor Grattan-Guinness, ed., *Landmark Writings in Western Mathematics, Case Studies 1640–1940* (Amsterdam: Elsevier, 2005), 292–302. I am deeply indebted to these works. Joan L. Richards, 'Historical Mathematics in the French Eighteenth Century', *Isis* 97 (2006), 700–13, is unfortunately marred by the unjustified attribution to Sylvestre François Lacroix of all the mathematical parts of volumes 3 and 4 of the 1799–1802 edition of the *Histoire*: 'In 1799 Sylvestre François Lacroix extended Montucla's historical narrative to include the analytic developments of the eighteenth century' (p. 702); 'the last two volumes of his second edition were primarily written by others. Most notable among these helpers was Sylvestre François Lacroix, who wrote the section on the development of analysis in the eighteenth century' (p. 707). As we will see below, after Montucla's death Lalande supervised the printing of volumes 3 and 4, and indeed asked the help of other scholars. However, Lacroix's contribution consisted only in revising article xxxiii on the integration of partial differential equations. 'Cet article étant un des plus difficiles de tout l'ouvrage, j'ai prié le cit. Lacroix, un de nos plus habiles géomètres, de vouloir bien le revoir': Jean E. Montucla, *Histoire des Mathématiques*, revised by Joseph Jérôme L. de Lalande (4 vols, Paris: Henri Agasse, 1799–1802; reprinted with a preface by Charles Naux, Paris: Blanchard, 1960), vol. 3, pp. 342–52; quote p. 342 (note).

support, contributed to Montucla's appointment as a foreign member of the Berlin Academy of Sciences in 1755.

The latter half of the eighteenth century witnessed an upsurge of interest in history as well as a refinement of historical criticism. Perhaps for the first time in European culture, human history came to be seen as progress, rather than decay from a Golden Age: progress leading from a state of ignorance and intolerance to an enlightened present of knowledge and civilization. Writings of the period, even writings of mathematicians such as d'Alembert, Lacroix, Joseph Louis Lagrange and Louis F.A. Arbogast, reflect this ideological interest in the progress of the human spirit, of which the mathematical sciences were held to be the greatest ornament. It is hardly surprising, therefore, to find so many articles in the *Encyclopédie* replete with historical information. Histories of astronomy such as those of Jean-Sylvain Bailly (1775–87) and Lalande were also composed in this period. Montucla was part of this cultural movement – in a way, he was its pioneer.[3]

The two volumes of Montucla's *Histoire* were published in 1758. They cover the history of mathematics from ancient times to the seventeenth century, the seventeenth century occupying the whole second volume. During his first years in Paris, Montucla made a living by working for the *Gazette*

---

3   Confining ourselves to French works we note: Jean Sylvain Bailly, *Histoire de l'astronomie ancienne* (1775); idem, *Histoire de l'astronomie moderne* (3 vols, 1779–1782); idem, *Traité de l'astronomie indienne et orientale* (1787); Jérôme Lalande, *Bibliographie astronomique* (1803); Jean Baptiste Joseph Delambre, *Rapport historique sur les progrès des sciences mathématiques depuis 1789* (1810). Lacroix wrote a lengthy historical preface in his *Traité du calcul différentiel et du calcul intégral* (1797–1800) and an 'Essai pour l'histoire des mathématiques pendant les dernières années du 18ème et le premier du 19ème' (MS at the Académie des Sciences, dossier personnel); Lacroix's historical work resonates with Montucla's treatment of the history of the calculus in volume 3 of the *Histoire*. One should also note the work, in several ways opposed to Montucla's, of Alexandre Savérien: *Dictionnaire universel de physique et de mathématiques* (1753); *Histoire critique du calcul des infiniment petits* (1753); *Histoire des progrès de l'esprit humain dans les sciences exactes* (1766). In the field of the history of medicine one should mention Daniel Le Clerc, *Histoire de la médicine* (1702).

de France. His family was not rich, yet he must have spent quite a lot of time in the unprofitable occupation of the erudite historian, for his books are extremely well-researched and show that a great deal of archival work was diligently executed by their author. In 1761 Montucla was appointed Secretary of the Intendance of Grenoble. As we shall see, Montucla led the quiet life of a public servant. However, there was at least one adventurous year in his life: between 1764 and 1765 he joined the disastrous mission to Cayenne organized by Étienne François Turgot (Jacques Turgot's brother), whose poor organization cost so many people's lives. Montucla had been promised a post as professor of hydrography in one of the naval schools upon his return. He was one of the recorded 918 people who survived (out of 10,000). In 1766 Montucla

> obtained a subordinate post in the superintendence of the Royal buildings, gardens, manufactures and academies, and moved to Versailles. He kept that post for twentyfive years, that is, until every bureau of the Royal administration was swept away by the triumphant revolution.[4]

The quiet appointments Montucla held under the *ancien régime* enabled him to focus on historical research. His works include a book on smallpox inoculation (1756), a new edition of Jacques Ozanam's *Récréations mathématiques et physiques* (1778), and a French translation of Jonathan Carver's *Travels through the Interior Parts of Northern America* (1784). While the Revolution left Montucla destitute, he must have had good contacts with the revolutionaries since in 1794 he received a national reward from the *Comité de salut public*. Then in 1795 he was ordered to examine the treaties deposited in the archives of the Ministry of Foreign Affairs, and in 1796 he was elected *associé non résident* of the mathematics division of the newly founded *Institut National*.

It was Lalande who, around this time, encouraged Montucla to publish a new edition of his *Histoire*, to include the eighteenth century. It is likely that Montucla had been pursuing this project for many years, since he soon presented his publisher, Henry Agasse, with a revised edition of

---

4   Sarton, 'Montucla', 524–5.

the 1758 *Histoire* (which was printed in 1799), as well as two additional tomes.[5] These were only printed posthumously (in 1802) and covered the eighteenth century, amounting to a total of 832 and 688 pages.[6]

As I have already mentioned, Montucla's *Histoire* is a product of the French Enlightenment. However, as wit is usually associated with this cultural movement, I hasten to say that it is hardly possible to find any witticism in the densely printed 3,000 pages that make up the four volumes of the *Histoire*. Montucla is verbose and shows a pedantic love for detail. The length of his work, however, is only partly due to pedantry. Montucla's view of what a history of mathematics should include is quite broad in chronological, geographical and thematic terms. As the subtitle reads, his history has the ambition of covering the 'progrès' of the mathematical sciences 'depuis leur origine jusqu'à nos jours'. Invoking the authority of Herodotus and Iamblichus, Montucla sets off by referring, with adequate scepticism, to the ancient Egyptian sage Hermes Trismegistus, whose inscriptions are said according to Ammianus Marcellinus to have inspired Pythagoras' interest in mathematical knowledge.[7] The first volume of the work covers the history of mathematics in Greek culture up to the fall of Constantinople, in the 'East' (this means 'Arabs, Persians, Chinese, and Hindus'), and in the Latin world up to the fifteenth century. The second volume is devoted to the seventeenth century. Montucla does not deal only with 'mathématiques pures', but also discusses mechanics (theoretical and practical), astronomy, optics, acoustics, and pneumatology.[8] As Sarton has noted, for its breadth

---

5   Ivor Grattan-Guinness informs us that 'Lalande stated that he had been pressing Montucla since 1792 to bring out these volumes, that printing had started in 1794, and that half of the third volume was in proof. Montucla was still alive at the time of Lalande's writing'. Ivor Grattan-Guinness, *Convolutions in French Mathematics, 1800–1840* (Basel: Birkhäuser, 1990), 142, referring to Jérôme Lalande, review of vols 1 and 2 of Montucla, *Histoire* (1799–1802), *Magasin encyclopédique* 3 (1799), 256–9.
6   The same year Bossut's more popular work appeared: Charles Bossut, *Essai sur l'histoire générale des mathématiques* (1802). It was translated into English by John Bonnycastle, into German by N.T. Reimer, and into Italian by A. Mozzoni.
7   Montucla, *Histoire* (1799–1802), 51.
8   See the *Système Figuré* on pp. xxvi–xxviii.

The Quarrel on the Invention of the Calculus

and scope Montucla's *Histoire* 'might almost be called a history of science written from the mathematical angle.'[9]

There is reason to believe that the third and fourth volumes of the *Histoire* were almost complete at the time of Montucla's death in December 1799.[10] More specifically, pages 1–336 of volume 3 of the new edition of the *Histoire* had already been proof-read and printed.[11] The rest was revised by Lalande, who availed himself of the help of several scholars: most notably, S.F. Lacroix revised pages 342–52 on the integration of partial differential equations. The priority dispute is tackled on pp. 102–19, therefore its narrative must be attributed entirely to Montucla.[12]

As is well known, the priority dispute between Leibniz and Newton was poisoned by nationalistic ideologies. No such chauvinism is discernible in Montucla, who rather tries his best to maintain an impartial point of view. Newton's priority (*anteriorité*)[13] cannot be denied, according to Montucla, and Leibniz is deemed responsible for having initiated the

---

9   Sarton, 'Montucla', 536.
10   Ibid., 552n.
11   'Note de l'Editeur. L'impression de cette feuille alloit finir lorsque l'auteur est mort, le 19 decembre 1799. La suite du manuscrit exige quelque révision et quelques additions dont je me suis chargé avec plaisir, comme un des plus anciens amis de Montucla, et comme ayant contribué beaucoup à lui faire entreprendre cette seconde édition. Jérome Lalande.' [Editor's note. The printing of this page was finished before the death of the author on 19 December 1799. The remainder of the manuscript required some revisions and additions, which I took on with pleasure, as one of Montucla's oldest friends and as one who had contributed much to his undertaking this second edition. Jérome Lalande.] Montucla, *Histoire* (1799–1802), vol. 3, p. 336n.
12   Montucla had already dealt with the priority dispute in the 1758 edition of his *Histoire*: Jean E. Montucla, *Histoire des Mathématiques* (Paris: Jombert, 1758), vol. 2, pp. 332–43. There are some repetitions in volume 3 of the 1799 edition, but also notable differences. In the first edition of the *Histoire* Montucla attributes the notes of the *Commercium epistolicum* to John Keill, whereas, as we will see below, in the second edition the notes are attributed to Newton. 'C'est-là la substance des apostilles de Keil au *Comm. Epistolicum*; mais elles sont toutes fausses ou du moins captieuses.' [This is the substance of Keill's annotations in the *Commercium epistolicum*; but they are all false, or at least misleading.] Montucla, *Histoire* (1758), vol. 2, p. 336.
13   Montucla, *Histoire* (1799–1802), vol. 3, p. 104.

dispute because of some letters he wrote to his English correspondents, in which he apparently claimed the invention of the calculus exclusively for himself. The 'journalistes de Leipsick' (the editors of the *Acta eruditorum*) are stigmatized for their unfair review of Newton's *De quadratura*.[14] Johann Bernoulli's criticisms of Newton's *Principia*, which played such a prominent role in the dispute, are discarded as 'injustes' since Newton's errors are the result of 'pure inadvertence'.[15] It should also be noted that in his third volume Montucla gives due credit to the contributions of British eighteenth-century mathematicians: for example, he informs his readers about the work on integration carried out by Roger Cotes, Colin Maclaurin and Thomas Simpson, and on infinite series by James Stirling and Brook Taylor. On the other hand, Newton and his acolytes are also criticized for the behaviour they adopted in the priority dispute. The most notable of these criticisms levelled against Newton concerns the pamphlet *Commercium epistolicum*, distributed in 1713. This presented the conclusions reached by the committee which had been appointed by the Royal Society to establish whether or not Leibniz had plagiarized the English mathematician.

Montucla states that he had learned from 'des notes de bonne main qui me sont venues d'Angleterre' that the footnotes to the *Commercium epistolicum* were not the result of the Committee's enquiries, but rather Newton's work.[16] This thesis, of course, was hardly new: it had long been circulating, along with other rumours about the priority dispute. The President of the Royal Society was suspected to be the person behind the *Commericum*. Still, it is remarkable that in his historical narrative Montucla is willing to confirm a version that throws such a bad light on Newton, a version that he derives from an informant I have so far been unable to identify.[17]

14 Ibid, 102.
15 Ibid, 105.
16 Ibid, 108.
17 Information on the authorship of the *Commercium epistolicum* certainly reached Montucla after the printing of the first edition of the *Histoire*, since, as we have remarked in note 12, in the first edition he attributed the notes to John Keill. Who Montucla's informant was is a mystery. Two candidates are Charles Hutton and

What Montucla has to say of the *Commericum* is rather interesting, since it is related to a controversy surrounding Georges-Louis Buffon's polemical endorsement of mathematical Newtonianism. Buffon is an example of how an Anglophile attitude could lead someone to take a polemical stand in favour of Newton. As Hanks and Roger have shown,[18] Buffon thrived in the intellectual climate of Dijon, where a colony of English aristocrats resided. His intellectual ties with Britain were numerous: he was a friend of the Duke of Kingston and Martin Folkes, found one of his collaborators in the microscopist John Turbeville Needham,[19] and was elected a Fellow of the Royal Society in 1739. In his youth, Buffon formed the idea of devoting himself to mathematics. Gabriel Cramer was among his correspondents. Even though Buffon later abandoned mathematics in order to focus on forestry and natural history, he has a place in the history of French mathematics as the translator – a rather free and inaccurate one – of Newton's *Method of Fluxions and Infinite Series*, which he knew in John Colson's English version (1736). *Méthode des fluxions et des suites infinies* appeared in 1740, prefaced by a few 'incendiary' pages[20] in which Buffon argued against the use of the infinite and of infinitesimals in mathematics, and discussed the priority dispute, siding with Newton against Leibniz: as Montucla later regretted. Buffon fully endorsed the theses upheld in the *Commercium epistolicum*, so Montucla set himself the task of disproving the footnotes and commentaries of the Royal Society pamphlet. Buffon's translation of Newton's tract on fluxions was part of a polemic he levelled against the kind of abstract mathematics carried out by mathematicians

---

Samuel Horsley. Hutton, as secretary of the Royal Society from 1779 to 1783, had access to the Royal Society's archives and corresponded with the French. Horsley, the editor of Newton's *Opera omnia* (1779–1785), was interested in mathematics, had access to many of Newton's mathematical manuscripts, and also corresponded with the French.

18 Lesley Hanks, *Buffon avant l'Histoire naturelle* (Paris: Presses Universitaires de France, 1966); Jacques Roger, trans. Sarah Lucille Bonnefoi, ed. L. Pearce Williams, *Buffon: a life in natural history* (Ithaca, NY and London: Cornell University Press, 1997).
19 Roger, *Buffon*, 41–3.
20 Hanks, *Buffon*, 112.

connected with the *Académie des Sciences*. Fontenelle, a defender of the legitimacy of the use infinitesimals, was one of his targets. Buffon did not approve of excessive reliance on symbolisms, and manifested his concern about their interpretation. He was especially concerned about the prevailing tendency he perceived to attribute a real denotation to symbols, such as Leibniz's differentials, that had none. The defenders of the use of infinity in mathematics, most notably Fontenelle, had affirmed the legitimacy of the mathematical infinite and infinitesimal, but according to Buffon the idea of infinity is an idea of privation, with no real object. So it is easy to understand why Newton's *Method of Fluxions*, based on kinematically representable fluent magnitudes, could been seen by Buffon as an alternative to the increasingly algebraized calculus practised by the Parisian mathematicians based at the *Académie*.

Montucla was instead quite sympathetic towards the mathematics practiced by some of the most eminent encyclopedists and *académiciens*, such as d'Alembert and Lagrange. *Contra* Buffon, who had endorsed the theses defended in the *Commercium*, he maintained not only that the notes of the *Commercium* were the prejudiced work of Newton, but also that they proved nothing concerning the priority dispute. In criticizing Buffon, Montucla revealed his own sympathies for the abstract mathematics favoured in the circle of encyclopedist mathematicians he was in touch with.[21]

Montucla's verdict on the so-called *Epistola posterior* that Newton sent Leibniz via Henry Oldenburg in October 1676 is exemplary in this respect. The *Epistola* was taken to be one of the main documents proving Leibniz's plagiarism. In the *Commercium*, the Committee (that is, Newton himself) claims that it was by reading this letter that Leibniz learnt the calculus. However, Montucla believes that the letter – the central piece

21 Savérien in *Histoire critique du calcul des infiniment petits* (1753) sided with Buffon, who he cited with approval (p. 20), and with Maclaurin's *Treatise of Fluxions* (1742) (pp. 7, 15, 36), where Fontenelle was criticized in a long footnote. Savérien was critical towards 'les Ouvrages des Calculateurs' (p. 35), epitomized by L'Hospital's *Analyse des Infiniment petits* (1696).

of evidence for Leibniz's plagiarism invoked in the *Commericum* – in fact proved nothing. He writes:

> Nous remarquons ici qu'après avoir lu et relu cette lettre, nous y trouvons seulement cette méthode [des fluxions] décrite, quant à ses effets et ses avantages, mais non quant à ses principes.
>
> [Here we note that, after having read and re-read this letter [the *Epistola posterior*], we only find the method of fluxions described as far as its consequences and advantages, but not as far as its principles.][22]

In the *Commercium*, Montucla continues, one can only find evidence of Newton's method of series, not of his method of fluxions. We can begin here to appreciate what Montucla meant by 'calculus' and by its 'discovery'.

For Montucla the calculus cannot be identified with techniques for series expansions (such as the binomial theorem), nor with techniques for the squaring of curves via infinite series, whereby one expands the integrand into a power series and integrates term-wise. The *Commercium* is all about these two topics.

So what is the calculus, according to Montucla? What techniques constitute the essence of Newton's and Leibniz's discovery? Montucla's narrative is based on two theses. The first is that very little effort had to be made by Newton and Leibniz in order to devise a notation and algorithm that had already been brought to a considerable state of perfection by Pierre de Fermat, Isaac Barrow, and John Wallis. Newton and Leibniz had just to systematize what was already available. The second thesis is that Newton, Leibniz and their followers advanced the integral calculus well beyond the techniques of their predecessors such as Wallis, Nicolaus Mercator, and James Gregory.

According to Montucla, it was Fermat, Barrow and Wallis who paved the way for the discovery of the notation and algorithm for the differential calculus.[23] Montucla even states that

---

22 Montucla, *Histoire* (1799–1802), vol. 3, p. 103.
23 Ibid, 109.

si l'on considère combien peu il y avoit à faire pour passer de leurs methodes au calculus différentiel; il paroîtra, ce semble, superflu de rechercher ailleurs l'origine de ce dernier

[if one considers how little was needed to pass from their methods to the differential calculus, it seems superfluous to search elsewhere for the origins of the latter].[24]

Montucla's view of the discovery of the differential calculus is thus continuist: he believes that Newton and Leibniz had been preceded by mathematicians who had achieved results that required only trivial additions in order to yield the calculus notation. From what Montucla states, it is clear that Newton's and Leibniz's contributions consisted in putting to good use the methods already devised by Fermat, Wallis and Barrow, whose notations and techniques they perfected.

When dealing with the integral calculus, Montucla stresses Wallis's contribution. He writes:

> Ajoutons, quant au calcul inverse, que Wallis avoit déjà désigné les élémens des aires des courbes par le rectangle fait de l'ordonnée et d'une portion infiniment petite de l'abscisse qu'il nommoit A, de sorte que l'élément de l'aire du cercle étoit, par example $A\sqrt{aa-xx}$. [...] A la caractéristique A de Wallis, substituez celle adoptée par Leibnitz, savoir dx pour la quantité x, voilà le calcul intégral.
>
> [As for the inverse calculus, let us add that Wallis had already defined the elements of the areas of curves by the rectangle formed by the ordinate and an infinitely small portion of the abscissa he calls A, so that the element of the area of the circle is found to be, for instance, $A\sqrt{aa-xx}$. [...] By replacing Wallis' characteristic A with that adopted by Leibniz, i.e. dx for quantity x, we get the integral calculus.][25]

After having stressed the continuity in methods between Newton and Leibniz and their predecessors, Montucla adds an important qualification:

---

24 Ibid.
25 Ibid.

> Mais un homme, autre que Leibnitz, peut être capable de faire ce pas, s'en seroit tenu là; au lieu que Leibnitz ne tarda pas d'appliquer l'un et l'autre calcul sous sa nouvelle forme aux problêmes les plus difficiles tant de la Géométrie que de la méchanique trascendante.
>
> [A man other than Leibniz, perhaps capable of taking such a step, would have gone no further; whereas Leibniz did not hesitate to apply both methods of calculus [i.e, the 'direct' differential calculus, and the 'inverse' integral calculus], in their new form, to the most difficult problems of geometry as well as transcendental mechanics.][26]

Montucla does not credit Newton and Leibniz with the invention of a new calculus or notation, or new fundamental concepts. In his narrative he rather conveys the idea that Newton and Leibniz's contribution consists in having furthered the application of a calculus, basically already achieved by Fermat, Wallis and Barrow, to difficult geometrical and mechanical problems. Most notably, what Montucla regards as a great step forward are Newton's and Leibniz's techniques of integration by substitution and by parts, techniques that go beyond the methods for squaring by infinite series expansion and infinite products already employed by Mercator, Wallis and James Gregory, and described in the *Commercium epistolicum*.

In this respect, it is important to consider how Montucla depicts the development of the integral calculus in the eighteenth century and how he evaluates the relative merits of the Newtonian and the Leibnizian schools. In dealing with eighteenth-century developments, Montucla gives pride of place to the integral calculus (devoting eighty pages of his third volume to the subject).[27]

In his narrative of the 'progrès' of the integral calculus Montucla gives credit to British mathematicians such as Taylor, Cotes, and Maclaurin. His verdict, though, is that the best results had been achieved by Continentals such as Johann Bernoulli, Leonhard Euler, Adrien Marie Legendre and Lagrange. Montucla here draws a distinction. He distinguishes between

---

26 Ibid, 109–10.
27 Ibid, 127–206. Montucla employs Louis Antoine de Bougainville, *Traité du calcul intégral* (1754) and Thomas Le Seur and François Jacquier, *Elemens du Calcul Intégral* (1768).

two 'principal cases' into which the integral calculus, or the 'inverse method of fluxions' can be divided.

> Le premier est celui des expressions où les quantités différentielles sont séparées les unes des autres [... le second cas est] celui qui fait sourtout le tourment des géomètres, est lorsque les variables sont mêlées avec leurs différentielles.
>
> [The first is that of expressions in which differential quantities are separate [...] the second – and the most troublesome one for geometers – is that in which variables are mixed with their differentials.][28]

The second, more difficult problem coincides with what we would call ordinary differential equations. In this case the Newtonians used power series developments, whereas the Continentals employed techniques that led to closed solutions. It is worth quoting Montucla's verdict on the Newtonian approach to integration via infinite series, compared to the one pursued by the Continentals:

> Le lecteur ne doit cependant pas en conclure que Neuton ait résolu le problème en entier; cela s'accorderoit mal avec ce qu'on dit plus haut. La méthode de Neuton donne seulement le rapport cherché en série infinie. Content de cette solution générale, Neuton n'a pas poussé plus loin ses recherches [...]. C'est pourquoi les Géomètres, réservant la méthode de Neuton pour les cas désespérés, ont recherché des moyens, soit pour intégrer en termes finis, lorsque cela se peut, soit pour séparer les indéterminées.
>
> [The reader should not conclude that Newton resolved the problem [of the integration of differential equations] completely; this would not fit well with what we have said above. Newton's method only delivers the sought relationship [between the independent and the dependent variables] as an infinite series [...]. This is why geometers, reserving Newton's method for the most desperate cases, have sought means, both for integrating in finite terms, when this is possible, and for separating the indeterminates.][29]

---

28   Montucla, *Histoire* (1799–1802), vol. 3, pp. 138–9.
29   Ibid., 165. The overall impression of a Continental superiority in the development of calculus techniques that emerges from the third volume of Montucla's *Histoire* was shared by some early nineteenth-century British authors, for instance by John

Montucla's *Histoire* is also notable for its lack of interest in issues relating to the 'rigour' of the calculus. Montucla does pay attention to famous quarrels such as those initiated by Michel Rolle at the *Académie des sciences* at the turn of the eighteenth century, or the polemic that followed the publication of George Berkeley's *Analyst* in 1734. Montucla's thesis, however, is that those who have questioned the foundations of the calculus are – like self-proclaimed circle-squarers or inventors of perpetual motion – nothing but 'ignorans' and 'esprits faux'.[30] Montucla praises Maclaurin's *Treatise of fluxions* (1742) as the work in which Newton's method was given rigorous form. Ultimately, though, he is dismissive of the Scottish mathematician. Maclaurin's demonstrations, he maintains, are rigorous but 'd'une longueur prodigieuse.' According to Montucla, Maclaurin should have confined himself to a few examples instead of writing a work that requires too much 'contention d'esprit.' Montucla believes that mathematicians should not embarrass themselves with questions concerning the nature of infinity that are addressed by the 'métaphisique la plus captieuse'.[31]

After this brief overview of Montucla's pages devoted to the polemic on the priority dispute and its aftermath, it is now time to attempt to define his views on the calculus and its development in the eighteenth century. It is these views that shape Montucla's narrative of the dispute. I would argue that Montucla's ideas reflect the position of mathematicians such as d'Alembert and Lagrange, and that they were given wide circulation through the mathematical articles of the *Encyclopédie*, which provided both mathematical and historical information. Montucla's *Histoire* was cited, and sometimes even copied *verbatim*, in many mathematical articles of the *Encyclopédie*. Montucla's association with the encyclopedists is thus well attested.

Toplis, the translator of Laplace's *Mécanique*, in his 'On the Decline of Mathematical Studies, and the Sciences dependent upon them', *Philosophical Magazine* 20 (1805), 25–31; and by John Playfair in 'Traité de Mechanique Celeste', *The Edinburgh Review* 22 (1808), 249–84.
30 Montucla, *Histoire* (1799–1802), vol. 3, p. 119.
31 Ibid.

Montucla's narrative of the calculus dispute is not nationalistic. It shows a moderate ecumenism between the Newtonian and Leibnizian schools. Such ecumenism is embedded in a book – the *Histoire* – that, as is typical of Enlightenment historiography, adopts comparative methods, espouses universalism, and is progress-oriented. For Enlightened *savants*, history has to reveal that the progress of knowledge is due to the collective cooperation of men and women of all cultures committed to seeking truth. In accordance with this view, Montucla refrained from attributing the discovery of the calculus to a single man or nation, but rather offered a nuanced view of how the calculus emerged thanks to contributions, both English and Continental, dating from the mid-seventeenth century. Montucla's *Histoire* is written very much in the style of Condorcet's *Tableau historique des progrès de l'esprit humain* (1795) or d'Alembert's *Discours préliminaire*.

In distancing himself from Buffon's partisan account of the priority dispute, Montucla sided with the *académiciens* whose abstract mathematics Buffon wished to criticize. Unlike Buffon, who praised the Newtonian version of the calculus, founded in terms of limits, as better grounded than the Leibnizian infinitesimalist one, Montucla showed little interest in questions concerning the foundations of the calculus. For him, advances in the applications of the calculus were more important than attempts to secure its foundations, and progress depended upon the development of algorithmic techniques, especially in the field of integration in closed form. This sensibility of Montucla's was deeply rooted in the Parisian mathematical community he was in contact with, and is very much reflected in his narrative of the priority dispute.

ADRIAN RICE
RANDOLPH-MACON COLLEGE, ASHLAND, VA

Vindicating Leibniz in the calculus priority dispute: The role of Augustus De Morgan

Introduction

It is today regarded as a matter of historical fact that Isaac Newton and Gottfried Wilhelm Leibniz both independently conceived and developed the system of mathematical algorithms known collectively by the name of calculus. But this has not always been the prevalent point of view. During the eighteenth century, and much of the nineteenth, Leibniz was viewed by British mathematicians as a devious plagiarist who had not just stolen crucial ideas from Newton, but had also tried to claim the credit for the invention of the subject itself.

Despite the fact that Newton was generally acknowledged to have developed his 'method of fluxions' several years previously, British writers alleged that 'ever since 1684, Leibnitz had been artfully working the world into an opinion, that he first invented this method'.[1] It was also maintained 'not only that Sir Isaac invented the method of Fluxions many years before Mr. Leibnitz knew any thing of it, but that Mr. Leibnitz took it from him'.[2] In contrast, the actions of Newton were portrayed as 'at all times dignified and just', especially when compared to the underhanded

---

[1] Charles Hutton, *A mathematical and philosophical dictionary* (2 vols, London: J. Johnson and G.G. & J. Robinson, 1795–6), vol. 2, p. 151.
[2] John Nichols, *Illustrations of the literary history of the eighteenth century, consisting of authentic memoirs and original letters of eminent persons*, vol. 4 (London: John Nichols, 1822), 4.

'pretensions of Leibnitz', who, it was claimed, was 'far inferior to Newton, both as a philosopher and as a man'.[3]

That British scientific writers were critical of (or simply downright hostile to) Leibniz's claims was partly due to the unrivalled position of Newton as the foremost mathematical scientist ever to have emerged from the British Isles. For well over a century after his death, Newton was almost universally regarded as a paragon of scientific, intellectual and moral virtue. And nowhere was this veneration more evident than in the land of his birth, where 'his glory [had] become that of the nation'.[4] Contemporary accounts referred to him as anything from 'that very excellent and divine theorist' to 'the most splendid genius that has yet adorned human nature', while his work was looked upon unambiguously 'as the production of a celestial intelligence rather than of a man'.[5]

One of the first British authors to challenge this viewpoint was the nineteenth-century mathematician Augustus De Morgan, who, in a series of works published between 1846 and 1855, set out to present Newton's life and achievements in a more human light. De Morgan was by no means the first to investigate the circumstances surrounding the dispute between Newton and Leibniz, nor would he be the last,[6] but his motivations differed con-

---

[3] David Brewster, *The life of Sir Isaac Newton* (London: John Murray, 1831), 217; Hutton, *Dictionary* (1795–96), vol. 1, p. 485; Thomas Thomson, *History of the Royal Society, from its institution to the end of the eighteenth century* (London: Robert Baldwin, 1812), 296.

[4] Charles Hutton, *A philosophical and mathematical dictionary* (2 vols, London: F.C. and J. Rivington, 1815), vol. 1, p. 525.

[5] Richard Bentley, *Eight Sermons, preached at the Hon. Robert Boyle's Lecture, in the year mdcxcii* (Oxford: Clarendon Press, 1809), 214; Thomson, *History*, 277; Hutton, *Dictionary* (1795–96), vol. 2, p. 150.

[6] The standard (and still authoritative) sources for information concerning the Newton–Leibniz calculus dispute are A. Rupert Hall, *Philosophers at war: The quarrel between Newton and Leibniz* (Cambridge: Cambridge University Press, 1980) and Derek Thomas Whiteside, *The mathematical papers of Isaac Newton* (8 vols, Cambridge: Cambridge University Press, 1967–1981), vol. 8, pp. 469–538. More recent studies include Skuli Sigurdsson, 'Equivalence, pragmatic platonism, and discovery of the calculus', in Mary Jo Nye, Joan L. Richards and Roger H. Stuewer, eds, *The invention*

siderably from those of his countrymen in that his prime objective was to set the historical record straight, regardless of considerations of nationality, religious affiliation or scientific renown. Or indeed of whom the outcome favoured. In this chapter, we examine De Morgan's research in this area and investigate the motivations that led him to initiate the rehabilitation of Leibniz among British mathematicians.

Newton vs. Leibniz

The story of the calculus priority dispute between Newton and Leibniz is very well known. However, for the purposes of this chapter, it will be useful to give a brief survey of the main events.[7]

It was Newton who created the first coherent system of calculus. During the period 1665–72, he developed and refined his method of fluxions, and undertook groundbreaking research in infinite series. Much of this work is contained in two lengthy papers, 'De analysi per æquationes numero terminorum infinitas' [On analysis by equations with an infinite number of terms] (1669) and 'Tractatus de methodis serierum et fluxionum'

---

of physical science: Intersections of mathematics, theology and natural philosophy since the seventeenth century (Dordrecht, Boston and London: Kluwer Academic Publishers, 1992), 97–116, which contrasts the historiographical positions of two late-nineteenth-century historians of mathematics, Moritz Cantor and Hieronymus Zeuthen, on the calculus priority dispute (esp. pp. 100–5); and Domenico Bertoloni Meli, *Equivalence and priority: Newton versus Leibniz, including Leibniz's unpublished manuscripts on the Principia* (Oxford: Clarendon Press, 1993), which discusses the rival theories of Newton and Leibniz in celestial mechanics, including reconsideration of issues of equivalence, publication styles and priority claims. See also the chapter in this volume by Niccolò Guicciardini on the treatment of the calculus controversy in Jean-Étienne Montucla and Joseph Lalande's *Histoire des mathématiques*.

7   An excellent one-paragraph summary of the controversy may be found in Niccolò Guicciardini, *The development of Newtonian calculus in Britain 1700–1800* (Cambridge: Cambridge University Press, 1989), 167–8.

[Treatise on the method of series and fluxions] (1671), both written during this period. Although Newton refrained from publishing these or any other calculus-related works at the time, he allowed manuscript copies to circulate among certain British colleagues, such as Isaac Barrow, John Wallis, John Collins and Henry Oldenburg.

Nearly a decade after Newton's first work on fluxions, another calculus system was developed, this time by the German mathematician and philosopher Gottfried Wilhelm Leibniz. Although he spent the majority of his life in what is now Germany, during the years 1672 to 1676 Leibniz was based in Paris, where he studied advanced mathematics under the guidance of one of the foremost European mathematical scientists of the time, Christiaan Huygens. No doubt also influenced in part by the earlier work of Bonaventura Cavalieri, Leibniz applied his extensive study of infinite series to problems in geometry, and in late 1675 brought these ideas together in a series of manuscripts, which formed the basis of what later became known as the differential and integral calculus.[8] By the time he left Paris in 1676, Leibniz was in possession of his own version of calculus, completely different from that of Newton both in terms of its notation and in the way it had developed. Yet, despite these key differences, the two methods were essentially equivalent in the results they could obtain and the problems they were able to solve.

Although Newton had created a fully developed set of calculus techniques by the early 1670s, it was Leibniz who was the first to disseminate his results to the scientific world at large by putting them into print. In 1684, he published a six-page paper in the newly founded journal *Acta eruditorum Lipsienium*. The paper, entitled 'Nova methodus pro maximis et minimis, itemque tangentibus, quae nec fractas, nec irrationales quantitates moratur, et singulare pro illis calculi genus' [A new method for maxima and minima as well as tangents, which is neither impeded by fractional nor irrational quantities, and a remarkable type of calculus for them], dealt purely with Leibniz's differential calculus. Two years later, he published another paper,

---

8   Joseph E. Hofmann, *Leibniz in Paris, 1672–76* (Cambridge: Cambridge University Press, 1974).

'De geometria recondita et analysi indivisibilium et infinitorum' [On recondite geometry and the analysis of indivisibles and infinities], in which his integral calculus appeared for the first time.[9]

Meanwhile, although tantalizing summaries of, and correspondence relating to, Newton's fluxional methods were published by John Wallis in his *Opera mathematica*,[10] it was not until 1704 that Newton's first publication on the fluxional calculus – *Tractatus de quadratura curvarum* (composed in the early 1690s) – finally appeared, and then only as an appendix to his book on *Opticks*. In the 'Advertisement' to the volume, Newton indicated the reason he had finally broken his long silence:

> In a Letter written to Mr. Leibnitz in the Year 1676, and published by Dr. Wallis, I mentioned a Method by which I had found some general Theorems about squaring Curvilinear Figures [...]. And some Years ago I lent out a Manuscript containing such Theorems, and having since met with some Things copied out of it, I have on this Occasion made it publick [...].[11]

Leibniz had made two brief visits to London, one in 1673 and one in 1676, on his way home from Paris. During this second stay, he had apparently seen a copy of the *De analysi* manuscript, containing Newton's methods of quadrature and infinite series. The implication here was clearly that Leibniz had obtained his methods from a perusal of Newton's work. (Of course, what Newton was unaware of was that Leibniz had already developed his own system by late 1675.)

Nevertheless, such a veiled accusation could not go unanswered, and in a review of the *Tractatus* published in the *Acta eruditorum* of January 1705, Leibniz responded in kind:

---

9   Gottfried Wilhelm Leibniz, 'Nova methodus pro maximus et minimus [...]', *Acta eruditorum* (October 1684), 467–73; idem, 'De geometria recondita et analysi indivisibilium et infinitorum [...]', *Acta eruditorum* (1686), 226–33.
10 John Wallis, *Opera mathematica* (3 vols, Oxford: Oxford University Press, 1693–1699), vol. 2, pp. 392–6; vol. 3, pp. 622–9, 634–45.
11 Isaac Newton, *Opticks: or a treatise of the reflexions, refractions, inflexions and colours of light* (London: Sam. Smith and Benj. Walford, 1704), 'Advertisement', ii.

cujus elementa ab inventore D. Godofredo Gullielmo Leibnitio in his actis sunt tradita
[...]. Pro differentiis igitur Leibnitianis D. Newtonus adhibet, semperque adhibuit, *fluxiones, quæ sunt quam proxime ut fluentium augmenta, æqualibus temporis particulis quam minimis genita*; iisque tam in suis *Principiis Naturæ Mathematicis*, tum in aliis postea editis, eleganter est usus; quem admodum et Honoratus Fabrius in suâ *Synopsi Geometricâ*, motuumque progressus Cavallerianæ methodo substituit.

[The elements of this calculus have been given to the public by its inventor Dr. Gottfried Wilhelm Leibniz [...]. Instead of the Leibnizian differences, then, Dr. Newton employs, and has always employed, *fluxions, which are almost the same as the increments of fluents generated in the least equal portions of time*. He has made elegant use of these both in his *Principia Mathematica* and in other later publications, just as Honoré Fabri, in his *Synopsis Geometrica* substituted the advance of motions for the method of Cavalieri.][12]

Leibniz's comparison of Newton to Honoré Fabri, who had notoriously plagiarized Cavalieri's work some years before, was clearly a huge affront, and, while no more justified than Newton's own implications, further impaired relations between the two sides.

The Newtonian camp replied in 1708 with a paper by the Scottish mathematician John Keill, who drew attention to

Hæc omnia sequuntur ex celebratissimâ nunc dierum Fluxionum Arithmeticâ, quam sine omni dubio primus invenit Dominus Newtonus, ut cui libet ejus Epistolas à Wallisio editas legenti, facile constabit; eadem tamen Arithmetica postea mutatis nomine & notationis modo à Domino Leibnitio in *Actis Eruditorum* edita est.

[that very celebrated arithmetic of fluxions which, without any doubt, Dr. Newton invented first, as can readily be proved by anyone who reads the letters about it published by Wallis; yet the same arithmetic afterwards, under a changed name and method of notation, was published by Dr. Leibnitz in the *Acta Eruditorum*.][13]

12   Gottfried Wilhelm Leibniz, anonymous review of Newton's 'Enumeratio linearum tertii ordinis' and 'Tractatus de quadratura curvarum', *Acta eruditorum* (January 1705), 30–6, at 34–5; see Isaac Newton, *Commercium epistolicum D. Johannis Collins, et aliorum de analysi promota* (London: Pearson, 1712), 108–9.

13   John Keill, 'Epistola ad clarissimum virum Edmundum Halleium Geometriae Professorem Savilianum, de legibus virium centripetarum', *Philosophical Transactions of the Royal Society* 26 (1708), 174–88, at 185.

Vindicating Leibniz in the calculus priority dispute 95

On seeing this paper, Leibniz wrote to the Royal Society, demanding that they arbitrate to obtain a retraction of Keill's charges of plagiarism. In response the Society appointed a committee to investigate the matter. After examining the relevant documents available to them (which, of course, largely supported Newton), the committee quickly presented their results in a short publication of 1712 entitled *Commercium epistolicum* [*Exchange of letters*], in which they concluded

> we reckon Mr. Newton the first Inventor; and are of Opinion, that Mr. Keill, in asserting the same, has been no ways injurious to Mr. Leibnitz.[14]

Leibniz had never denied that Newton's calculus predated his. What he wanted was a vindication from the charge of plagiarism and an acknowledgement that his work was created independently. That this was denied him is hardly surprising for two reasons. Firstly, this supposedly impartial committee never bothered to solicit any testimony from Leibniz; and secondly, both the anonymous author of the committee's report and the President of the Royal Society, to whom the committee was ultimately responsible, were one and the same person: Newton himself.[15]

Not surprisingly, the argument continued for some time, becoming increasingly bitter. Even Leibniz's death in 1716 could not completely halt the recriminations. So resentful had Newton become that in 1722, at the age of nearly eighty, he personally superintended a second edition of the *Commercium epistolicum*, providing an anonymous preface, as well as altering the original text to harden the language and strengthen the conclusions against Leibniz.[16] From then on, in Britain at least, Leibniz was regarded among British mathematicians as a deceitful schemer who had plagiarized Newton's work on the calculus and attempted to claim priority for its invention. And this attitude had not changed in any significant way by the early nineteenth century.

---

14   Newton, *Commercium epistolicum* (1712), 122.
15   Hall, *Philosophers at War*, 178; Whiteside, *Mathematical Papers*, vol. 8, pp. 539–60.
16   Isaac Newton, *Commercium epistolicum D. Johannis Collins, et aliorum de analysi promota* (second edition, London: J. Tonson and J. Watts, 1722).

Early nineteenth-century British Newtonian studies

By contrast, the unadulterated praise of Isaac Newton continued unabashed. As De Morgan later wrote, 'it was in Britain the temper of the age [...] to take for granted that Newton was human perfection'.[17] This was in part symptomatic of the attitude towards biography at this time, when it was considered the duty of biographers to highlight the honour and ethical strengths of their subjects as much as to provide a dispassionate assessment of their achievements. Thus, in the *Life of Sir Isaac Newton*, published in 1831 by the Scottish physicist Sir David Brewster, abundant praise of 'the most exalted genius'[18] was conspicuous throughout.

Not surprisingly, on the matter of the calculus priority dispute, Brewster's opinions were no different from those of his predecessors: Newton was the inventor and Leibniz the plagiarist. Whereas Brewster regarded Newton's behaviour in the affair as 'at all times dignified and just', he alleged that 'the conduct of Leibniz was not marked with the same noble lineaments'.[19] As far as Brewster was concerned, Newton's actions required no explanation, since 'he knew his place as a philosopher, and was determined to assert and vindicate his rights.'[20] Paradoxically however, by attempting to defend the same privileges, Leibniz had 'cast a blot upon his name, which all his talents as a philosopher, and all his virtues as a man, will never be able to efface'.[21]

---

17   Augustus De Morgan, ed. Sophia Elizabeth De Morgan and Arthur Cowper Ranyard, *Newton: his friend and his niece* (London: Elliot Stock, 1885) (excerpts reprinted in Rob Iliffe, Milo Keynes and Rebekah Higgitt, eds, *Early biographies of Isaac Newton, 1660–1885* (2 vols, London: Pickering & Chatto, 2006), vol. 2, pp. 289–337), 130.
18   Brewster, *Life*, 2.
19   Ibid., 217, 218.
20   Ibid., 338.
21   Ibid., 218.

Figure 2: De Morgan's idiosyncratic cartoon depiction of the Newton–Flamsteed dispute, inserted in A. De Morgan, 'Mathematical Biography extracted from the Gallery of Portraits', Royal Astronomical Society Archives, De Morgan MSS. (Copyright © Royal Astronomical Society.)

Just four years after Brewster's biography, however, the first English work to cast aspersions on Newton's integrity appeared. Written by the astronomer Francis Baily, this was a substantial biography of John Flamsteed,[22] the first Astronomer Royal, who, amongst other things, had provided important astronomical data for Newton's use in the *Principia*. Through Baily's discovery of previously unpublished letters and manuscripts, it was revealed that Newton had treated Flamsteed unfairly over the publication of the latter's star catalogue, and had even deleted references to him in later editions of the *Principia*.[23] Baily's *Account of the Rev. John Flamsteed* deeply surprised and unsettled the British scientific community, not just because he had dared to expose a blemish in Newton's character, but also by the extensive array of documents he had employed to support his argument. The first blow had been struck at the image of Newton as the epitome of moral rectitude.

Baily's revelations about Newton's personality were deeply disturbing to many in the British scientific community. In addition to David

---

22 Francis Baily, *An account of the Rev. John Flamsteed* (London: William Clowes, 1835).
23 For a summary of the Newton–Flamsteed dispute, see Richard S. Westfall, *Never at rest: A biography of Isaac Newton* (Cambridge: Cambridge University Press, 1980), 541–50, 655–66; also consult David H. Clark and Stephen P.H. Clark, *Newton's Tyranny: The Suppressed Scientific Discoveries of Stephen Gray and John Flamsteed* (New York: W.H. Freeman and Co., 2001).

Brewster, others who were also troubled by Baily's disclosures were the Oxford astronomer Stephen Peter Rigaud and the Cambridge natural philosopher William Whewell. As recent historians of science have argued, for men such as these, unsettling discoveries about Newton damaged their idea of him as an unequivocal model of national, scientific and religious morality. This is not altogether surprising when it is observed that 'one of the most distinctive features of British intellectual life in the eighteenth century, and in much of the nineteenth, was the extent to which science was seen to be allied to the cause of religion'.[24] This view is corroborated by Richard Yeo:

> There is evidence from the debates of the 1830s that this was not merely a secondary issue but one which contained serious implications for assumptions about science. [...] Pleading with Whewell to enter the debate, Rigaud said that 'if Newton's character is lowered, the character of England is lowered, and the cause of Religion is injured'.[25]

To play his part in Newton's defence, Rigaud published two works containing relevant and supportive archival material. The first, a short *Historical essay on the first publication of Sir Isaac Newton's Principia*, uncovered new information about Edmond Halley, along with portions of his correspondence with Newton, while the second, the lengthy *Correspondence of scientific men of the seventeenth century*,[26] consisted of primary source material from 1606 to 1742 and included Newton, Flamsteed, Halley, Roger Cotes, Isaac Barrow, John Collins and John Wallis among the correspondents.

---

24 John Gascoigne, 'From Bentley to the Victorians: The rise and fall of British Newtonian natural theology', *Science in context* 2 (1988), 219–56, at 219.
25 Richard Yeo, 'Genius, method, and morality: Images of Newton in Britain, 1760–1860', *Science in Context* 2 (1988), 257–84, at 271.
26 Stephen Peter Rigaud, *Historical Essay on the first publication of Sir Isaac Newton's Principia* (Oxford: Oxford University Press, 1838); Stephen Peter Rigaud and Stephen Jordan Rigaud, *Correspondence of scientific men of the seventeenth century* (2 vols, Oxford: Oxford University Press, 1841).

Whewell's contribution was his three-volume *History of the inductive sciences* of 1837, in which an underlying theme is the exemplary moral and intellectual qualities of great scientists and philosophers. In this vein Newton, a greater genius than Flamsteed, is consequently more worthy of our attention and sympathy, since Flamsteed, 'though a good observer, was no philosopher [...] and was incapable of comprehending the object of Newton's theory'.[27] 'Thus,' as Geoffrey Cantor put it,

> not only was Newton amply endowed with the apposite intellectual qualities, but he was also, in Whewell's opinion, morally virtuous – 'candid and humble, mild and good'.[28]

It was at this point that Augustus De Morgan entered the discussion.

De Morgan's early contributions to the debate

Although best remembered today for the famous laws which bear his name,[29] Augustus De Morgan is otherwise largely unknown to the majority of today's mathematicians. Born to British parents in India in 1806, De Morgan was educated at Trinity College, Cambridge, from 1823 to 1827, where he came under the influence of tutors such as William Whewell,

---

27 William Whewell, *History of the inductive sciences* (3 vols, London: J.W. Parker, 1837), vol. 2, p. 198.
28 Geoffrey N. Cantor, 'Between rationalism and romanticism: Whewell's historiography of the inductive sciences', in Menachem Fisch and Simon Schaffer, eds, *William Whewell: A composite portrait* (Oxford: Clarendon Press, 1991), 67–86, at 80.
29 De Morgan's Laws are usually encountered by students in logic, where they can be stated as:
$$\neg(a \vee b) = \neg a \wedge \neg b \text{ and } \neg(a \wedge b) = \neg a \vee \neg b$$
or in set theory, where they are often given as:
$$(A \cup B)' = A' \cap B' \text{ and } (A \cap B)' = A' \cup B'.$$

the algebraist George Peacock and the future Astronomer Royal George Biddell Airy. Although he graduated from Cambridge as fourth wrangler, his nonconformist religious beliefs led him to forego an almost guaranteed college fellowship; instead he was appointed the founding professor of mathematics at the non-denominational University College London (UCL) at the age of only 21.[30] He resigned his chair in 1831, only to return in 1836 for thirty more years before resigning again in 1866. He remains the only professor at UCL to have resigned *twice* – both times on matters of principle.[31]

An intriguing character whose high intellect was equalled by a sharp wit and keen sense of humour, De Morgan wrote on almost every aspect of pure mathematics, contributing particularly to British work in symbolic logic and abstract algebra. His status as one of the most respected and influential mathematicians of mid-nineteenth-century Britain was reflected in his tenure as the first president of the London Mathematical Society, founded in 1865.

In addition to his work in mathematics and logic, De Morgan had a lifelong fascination for the history of mathematics, being 'perhaps more deeply read in the philosophy and history of mathematics than any of his contemporaries'.[32] He contributed over 700 articles to the well-respected *Penny Cyclopædia* on all areas of mathematical science, including one in which he invented the term, although not the method, of 'mathematical induction'. His extensive historical erudition was fuelled by a passionate bibliophilism, through which his library grew through a period of 40 years to stand at well over 3000 mathematical volumes by the time of his death

---

30   Adrian Rice, 'Inspiration or Desperation? Augustus De Morgan's appointment to the Chair of Mathematics at London University in 1828', *British Journal for the History of Science* 30 (1997), 257–74.
31   His first resignation was occasioned by the dismissal of a professorial colleague. He resigned in 1866 because the College refused to appoint a professor on the grounds of his being a controversial Unitarian minister, a decision De Morgan regarded as a violation of its founding principle of religious neutrality.
32   Walter William Rouse Ball, *A history of the study of mathematics at Cambridge* (Cambridge: Cambridge University Press, 1889), 133.

in 1871.³³ Thus, his mathematical expertise, strong interest in history and wide knowledge of the literature, coupled with his nonconformist tendencies and high moral principles made him an apt scholar to contribute to the Newton–Leibniz debate.³⁴

De Morgan's first contribution to the subject was a 40-page biography of Newton, published in 1846, which was, 'after Baily's *Life of Flamsteed*, the first English work in which the weak side of Newton's character was made known'.³⁵ While still acknowledging Newton's greatness and intellectual achievements, De Morgan drew considerable attention to Newton's imperfections, claiming that

> [t]he great fault, or rather misfortune, of Newton's character was one of temperament [... which] showed itself in fear of opposition: when he became king of the world of science it made him desire to be an absolute monarch; and never did monarch find more obsequious subjects. His treatment of Leibnitz, of Flamsteed, and (we believe) of Whiston is, in each case, a stain upon his memory.³⁶

---

33   For more on De Morgan's life, see Sophia Elizabeth De Morgan, *Memoir of Augustus De Morgan* (London: Longmans, Green & Co., 1882).

34   On De Morgan as historian, see Joan L. Richards, 'Augustus De Morgan, the history of mathematics, and the foundations of algebra', *Isis* 78 (1987), 7–30; Adrian Rice, 'Augustus De Morgan: historian of science', *History of science* 34 (1996), 201–40; and Rebekah Higgitt, *Recreating Newton: Newtonian biography and the making of nineteenth-century history of science* (London: Pickering & Chatto, 2007), chapters 4–6.

35   S.E. De Morgan, *Memoir*, 256.

36   Augustus De Morgan, 'Newton', in *The Cabinet portrait gallery of British worthies*, vol. 11 (London: Charles Knight & Co., 1846), 78–117 (reprinted in Augustus De Morgan, ed. Philip E.B. Jourdain, *Essays on the life and work of Newton* (Chicago and London: The Open Court Publishing Company, 1914), 3–63), at 98, 100. William Whiston was Newton's successor as Lucasian Professor of Mathematics at Cambridge in 1703. He was removed from his chair in 1710 on the grounds of his Arian faith; particularly ironic since his predecessor had similar (but secret) unorthodox religious beliefs. In 1720, Newton used his position as President of the Royal Society to block a proposal to make him a Fellow.

De Morgan's revisionist attitude, fuelled by the revelations about Newton's behaviour towards Flamsteed, led to a natural curiosity about his dealings with Leibniz. This resulted in the composition of a series of articles regarding the calculus controversy and, in particular, the *Commercium epistolicum*. Ironically perhaps, De Morgan's first paper on the subject was actually on a point that favoured Newton. According to all English published sources to date, the committee which had produced the *Commercium epistolicum* consisted of six men, all British. However, in a letter published in Joseph Raphson's *History of fluxions*, Newton had claimed that evidence was 'collected and published by a *numerous* Committee of gentlemen *of different nations*',[37] thus creating the misleading impression that the committee was impartial.

However, after conducting his own research on the subject, including a detailed inspection of the Royal Society minute books, De Morgan discovered that four extra people, including two non-Britons, were later added to the committee. These additions went unrecorded in print, except in an obscure eighteenth-century French work in De Morgan's possession, which revealed Abraham de Moivre to be one of the two foreigners selected. But it was the book's assertion that the appointment 'drew De Moivre out of the neutrality which till then he had observed',[38] that particularly intrigued De Morgan, since the implication was clearly that simply joining the committee was tantamount to de Moivre taking sides in the dispute. His conclusion could only be 'that the Committee in question was thought at the time not to be a judicial body, but one of avowed partizans'.[39]

37 Joseph Raphson, *The history of fluxions* (London: William Pearson, 1715); Augustus De Morgan, 'On a point connected with the dispute between Keill and Leibnitz about the invention of fluxions', *Philosophical Transactions of the Royal Society* 136 (1846), 107–9, at 107; Isaac Newton, 'An account of the book entitled *Commercium epistolicum* [...]', *Philosophical Transactions of the Royal Society* 29 (1715), 173–224, at 221.
38 De Morgan, 'On a point', 108.
39 Ibid.

Realising the significance of this discovery, De Morgan submitted a short paper on the subject to the Royal Society, intending it to appear as a brief note in its *Proceedings*. However, 'To my very great surprise, they were printed in all the dignity of the *Philosophical Transactions*, in which no historical paper has ever appeared, that I know of – certainly none within the century.'[40] Encouraged by this positive response to his research, he drew up another paper, entitled 'A comparison of the first and second editions of the Commercium Epistolicum'. This paper, while clearly related to the subject of the first, cleared up a point in favour of Leibniz. Once again, De Morgan submitted it to the Royal Society on the grounds that, as he explained in his manuscript, 'the Royal Society was made the instrument by which a signal injustice was perpetrated'.[41] He thus felt that 'the memory of Leibnitz has a peculiar claim upon that body for reparation of many wrongs'.[42]

His paper identified more than twenty undeclared additions and alterations to the second edition of the *Commercium epistolicum*.[43] Although not all of the changes were significant, De Morgan observed that 'the general tendency of the additions to bring out the unfairness of the original, and to convert hints into assertions, is curiously exemplified'.[44] The overall effect of the additions was to strengthen the language of the report against Leibniz, amounting to 'the falsification of a record in a matter affecting [Leibniz's] character, done under the name of the [Royal] Society'.[45] Once again, being 'unfortunate enough to differ from the general opinion in England as to

---

40   De Morgan, *Newton: his friend and his niece*, 132.
41   Augustus De Morgan, 'A comparison of the first and second editions of the Commercium Epistolicum' [1847], London, Royal Society Archives, AP/29/2, fols 1–12, at fol. 11.
42   Augustus De Morgan, 'On the additions made to the second edition of the Commercium Epistolicum', *Philosophical magazine* (third series) 32 (1848), 446–56, at 447.
43   Whiteside notes that '[De Morgan] does not seem to suspect that the agent was Newton': *Mathematical Papers*, vol. 8, p. 486, note 57.
44   De Morgan, 'On the additions', 450.
45   Ibid., 447.

the manner in which Leibnitz was treated' it fell to De Morgan to redress the historical balance.[46]

This was not a view that was fully shared by the Royal Society. In the opinion of the referee (whose confidential report De Morgan did not see), 'if the question of repairing a wrong done 140 years ago be entertained, it must be entertained in a much more formal & solemn manner' than by the publication of a single paper. If justice to Leibniz were to be done, he said, 'it should be done by a committee appointed for that purpose'.[47] Somewhat confusingly however, the Society also stated its belief that 'it is no peculiar duty of the Council of today to vindicate or to blame the proceedings of their predecessors'.[48] But despite these contradictory views, De Morgan's paper was rejected, a decision which annoyed him intensely and to which he was to refer in more than one of his later works: 'I freely and unreservedly blame the Council of the Royal Society [...] for not printing the account of the variations mentioned above,' he wrote, adding, 'they missed a golden opportunity'.[49]

---

46 De Morgan, 'On a point', 108. In 1856, fully aware of and in agreement with De Morgan's work, Jean-Baptiste Biot and Félix Lefort published their annotated French edition of the *Commercium*: Jean-Baptiste Biot and Felix Lefort, eds, *Commercium epistolicum* (Paris: Mallet-Bachelier, 1856). But, as Rouse Ball commented somewhat sarcastically a generation later, although 'these writers agree in saying that it shews a marked bias in favour of Newton, [...] it hardly needs so much labour to prove that the report of a committee which only heard one side is not impartial': Walter William Rouse Ball, *A short account of the history of mathematics* (London: Macmillan, 1888), 329–30.

47 George Peacock, letter to Samuel Hunter Christie, 16 October 1847: London, Royal Society Archives, RR/1/57.

48 Samuel Hunter Christie, letter to Augustus De Morgan, 25 February 1848. Pasted in the copy of Newton, *Commercium* at University of London, Senate House Library, shelfmark [De M] L [B.P.1], Secondary Strong Room.

49 De Morgan, *Newton: his friend and his niece*, 135. Throughout his entire career, and despite being on very friendly terms with several fellows, De Morgan consistently refused to be nominated as a Fellow of the Royal Society. See Rebekah Higgitt, 'Why I don't FRS my tail: Augustus De Morgan and the Royal Society', *Notes and records of the Royal Society* 60 (2006), 253–9.

De Morgan vs. Brewster

It was four years before De Morgan re-entered the debate but, he said, he had never 'come fresh to this controversy of Newton and Leibnitz without finding new evidence of the atrocious unfairness of the contemporary partisans of Newton'.[50] And prompted by the publication of the correspondence between Newton and Roger Cotes and, more significantly, of Leibniz's mathematical manuscripts,[51] he returned to the subject in 1852. His work was the first English analysis of the priority dispute since the discovery of Leibniz's independent development of the calculus found among his papers, which included 'various original drafts, containing problems in which both the differential and integral calculus are employed'.[52] Here was the long-awaited proof that Leibniz had not plagiarized the work of Newton but had created his calculus separately and independently.

Up to this point it had always been believed that it was solely Newton's followers who had brought the charges of plagiarism, with Newton himself staying aloof from the sordid wranglings. However, in another paper of 1852, De Morgan called even this into question, challenging Newton's integrity still further. He claimed that the anonymous author of a review of the *Commercium epistolicum* in the *Philosophical Transactions*[53] and the Latin preface to the second edition of the *Commercium* was none other than Newton himself, a hypothesis that remained unproved until 1855, when confirmation was to come from a most unlikely source.

David Brewster's epic *Memoirs of the life, writings, and discoveries of Sir Isaac Newton* (1855) was the work of nearly two decades. Initially provoked

---

50 Augustus De Morgan, 'A short account of some recent discoveries in England and Germany relative to the controversy on the invention of fluxions', *Companion to the almanac for 1852*, 5–20 (reprinted in De Morgan, *Essays*, 67–101), at 8.
51 J. Edleston, *Correspondence of Sir Isaac Newton and Professor Cotes* (London: John W. Parker, 1850); Karl Immanuel Gerhardt, ed., *G.W. Leibniz mathematische schriften* (7 vols, Halle: H.W. Schmidt, 1849–1863; reprinted Hildesheim: Georg Olms, 1962).
52 De Morgan, 'A short account', 17–18.
53 Newton, 'An account of the book'.

by Baily's revelations regarding Flamsteed, its length was more than doubled, firstly by Brewster's use of newly available manuscript sources, and secondly by the increasing necessity to defend Newton from the growing charges against him. During the course of its preparation De Morgan had been consulted on various points even though, as Brewster admitted, 'on a few questions in the life of Newton, and the history of his discoveries, my opinion differs somewhat from his'.[54] But despite this, Brewster verified 'from the documents in my possession, many of his views on important points which he was the first to investigate and to publish'.[55]

One of these points was the authorship of the anonymous 1715 review and 1722 preface to the *Commercium epistolicum*. Through his research, Brewster was able to confirm for the first time, via information derived from unpublished documents, De Morgan's conjecture concerning the writer of these two pieces. 'Professor De Morgan,' he stated,

> had made it highly probable that both the review and the preface were written by Newton. Of the correctness of this opinion I have found ample evidence in the manuscripts [...] and it is due to historical truth to state that Newton supplied all the materials for the *Commercium Epistolicum*, and that [...] Newton was virtually responsible for its contents.[56]

But despite De Morgan's research, plus the mass of new evidence available, Brewster's new biography, like his earlier work, stopped little short of hero worship. Although without question the most comprehensive biography of Newton to date, it was still a very biased account, Brewster's partiality being particularly evident in his discussion of the calculus priority dispute. Although he recognised some of De Morgan's findings, he completely disregarded many of those favourable to Leibniz. Indeed, it is difficult to believe that Brewster had read the works of De Morgan which he cites:

---

54 David Brewster, *Memoirs of the life, writings, and discoveries of Sir Isaac Newton* (2 vols, Edinburgh: Thomas Constable & Co., 1855), vol. 1, p. xiii.
55 Ibid., xiii.
56 Ibid., vol. 2, p. 75. It was not until over a century later that Whiteside was able to prove categorically that Newton was the *de facto* author of the *Commercium epistolicum* itself; see Whiteside, *Mathematical Papers*, vol. 8, pp. 539–60.

his conclusion on Leibniz reads, word for word, identically to his judgement of 1831.[57]

Such blatant disregard for historical accuracy could not go unanswered. Immediately prior to its publication, De Morgan had been commissioned to write a review of the book for the Edinburgh-based *North British Review*. He had already stated his view, in a letter to former Lord Chancellor and fellow Newton aficionado Henry Brougham, that Brewster 'has done very good service, though he is a partisan'.[58] His critique was a thirty-page extension of that opinion. He stopped short of echoing Whewell's belief that such a monumental biography would be better undertaken by 'some person not so onesided and rhetorical as Sir David', but agreed that Brewster was 'still too much of a biographer, and too little of an historian' for such a project.[59]

On the calculus controversy his opinions are direct and unequivocal. 'I have no doubt,' he wrote in a letter to Brewster,

> that Leibnitz was used by the English with every sort of unfairness, and that Newton was a party to the ill usage [...]. I think that the attempt to show that it was *possible* that Leibnitz could have got any hint from what he saw of Newton's was a piece of effrontery. I think that Newton himself acted in a manner not becoming a gentleman in several particulars.[60]

In his review he continued:

---

57   Brewster, *Life*, 218; Brewster, *Memoirs*, vol. 2, p. 83.
58   Augustus De Morgan, Letter to Lord Brougham, 16 May 1855: University College London Archives: Brougham Correspondence, no. 10,299.
59   William Whewell, letter to Augustus De Morgan, 2 September 1856: University of London, Senate House Library, MS 775/370/34/1. Augustus De Morgan, review of Brewster's *Memoirs of the life of Sir Isaac Newton*, North British Review 23 (1855), 307–38 (reprinted in De Morgan, *Essays*, 119–82), at 310. For a more recent view of Brewster as a historian, see John R.R. Christie, 'Sir David Brewster as an historian of science', in A.D. Morrison-Low and John R.R. Christie, eds, *Martyr of science: Sir David Brewster 1781–1868* (Edinburgh: Royal Scottish Museum, 1984), 53–6.
60   Augustus De Morgan, letter to David. Brewster, n.d.: Royal Astronomical Society Archives, De Morgan MSS.

> We shall not stop to investigate the various new forms in which Sir D. Brewster tries to make [Leibniz] out tricking and paltry. We have gone through all the stages which a reader of English works can go through. We were taught, even in boyhood, that the Royal Society had made it clear that Leibnitz stole his method from Newton. By our own unassisted research into original documents we have arrived at the conclusion that he was honest, candid, unsuspecting, and benevolent.[61]

It is here, at the culmination of ten years of intense interest and immersion in the details of the calculus dispute, that we can truly feel what Whiteside later described as De Morgan's 'puritanical wrath'[62] at the treatment of Leibniz by Newton and his supporters. Although De Morgan's opinion of Leibniz's character (probably somewhat exaggerated for polemical purposes) is almost as flawed as the romanticized vision of Newton he so vigorously opposed, his sense of anger and frustration at the sheer unfairness of the affair is palpable.

It was not only in their opinions that Brewster and De Morgan differed. In their attempts to redress the historical balance, they emphasized two very different aspects of Newton's character. Indeed, as Paul Theerman persuasively argues, perusal of the two men's biographies tells us as much about the authors' personalities as that of their mutual subject.

> Brewster was characteristically concerned with public priority and position, De Morgan with private morals [...]. Brewster, a more public man concerned with promoting a more prominent status for the scientist, meshed his interpretation with the prevailing political ideas of the 'man of capacity'. De Morgan evoked instead an image of the solitary and perhaps eccentric scholar.[63]

The apparent obsession with the question of Newton's moral character, shown especially by Brewster and De Morgan, as well as by William Whewell, has itself to be understood in its historical context. As we have seen, Whewell in particular drew strong links between moral character and

---

61 De Morgan, review of Brewster, 321.
62 Whiteside, *Mathematical Papers*, vol. 8, p. 486.
63 Theerman, Paul, 'Unaccustomed role: The scientist as historical biographer – two nineteenth-century portrayals of Newton', *Biography* 8 (1985), 145–62, at 146, 158–9.

high intellect, a belief also firmly held by Brewster. However, as Richard Yeo has demonstrated,

> by the middle of the nineteenth century, Newton was a far more complex figure than the celestial or divine genius lauded by his contemporaries. While still exalted as the apex of scientific achievement, his genius was seen as more human in kind, and could not be dissociated from the evidence of passions, lapses, and delusions which cast severe doubts on the previous convictions about the affinity of intellectual and moral virtue.[64]

But although Yeo claims that 'neither Whewell, Brewster, nor De Morgan was able to reinstate unequivocally the alliance between intellectual and moral virtue',[65] and although this programme was indeed essential to established church members like Brewster and Whewell, De Morgan's religious nonconformity removed this objective from his agenda entirely. As he said:

> The scientific fame of Newton [...] gave birth to the desirable myth that his goodness was paralleled only by his intellect. That unvarying dignity of mind is the necessary concomitant of great power of thought is a pleasant creed, but hardly attainable [...]. [W]e live in discriminating days, which insist on the distinction between intellect and morals.[66]

---

64 Yeo, 'Genius, method, and morality', 278–9.
65 Yeo, Richard, *Defining science: William Whewell, natural knowledge, and public debate in early Victorian Britain* (Cambridge: Cambridge University Press, 1993), 144.
66 De Morgan, *Newton: his friend and his niece*, 140. Margaret Osler rightly comments, 'De Morgan's criticism of Brewster must, however, be tempered, for in his own book [i.e. *Newton: his friend and his niece*] he bent over backwards to clear Newton's moral reputation against a background of incriminating evidence': Margaret J. Osler, 'A hero for their times: early biographies of Newton', *Notes and records of the Royal Society* 60 (2006), 291–305, at 301. She is referring here to the extensive research undertaken by De Morgan on the question of the marital status of Newton's niece, Catherine Barton, and his friend and patron Charles Montague, later the Earl of Halifax. Since the propriety of their relationship was never well-defined publicly, this unresolved issue would have been deeply unsettling to Victorians like De Morgan.

De Morgan's secularity thus allowed him to investigate the question of Newton's morals for its own sake, without reference to scientific eminence or religious affiliation, which was, to him, immaterial. He therefore distinguished himself from his contemporaries by attempting, in the words of Joan Richards, 'to understand his mathematical predecessors not merely as intellectual forefathers but as human beings'.[67]

Thus, however 'scathing and irreverent'[68] his words may have seemed, his admiration and respect for Newton's intellectual accomplishments remained unchanged; indeed, there is every reason to believe that his fascination with Newton the man actually increased. As he wrote:

> Newton always right, and all who say otherwise excathedrally reproved is a case for ostracism [...]. But Newton of whom wrong may be admitted, Newton who must be defended like other men, and who cannot always be defended, is a man in whom to feel interest even if we are obliged to dissent from his eulogist.[69]

In contrast, then, to the heroic depiction given by Brewster, De Morgan's study of Newton was a more realistic representation, daring to highlight and criticize the character flaws and erratic behaviour, while in no way denigrating the towering achievements for which he is famous. In this way, De Morgan helped to initiate a more complex and human view of Newton that has dominated Newtonian studies ever since.[70]

---

67   Richards, 'Augustus De Morgan', 17.
68   Niccolò Guicciardini, '"Gigantic implements of war": images of Newton as a mathematician', in Eleanor Robson and Jacqueline Stedall, eds, *The Oxford handbook of the history of mathematics* (Oxford: Oxford University Press, 2009), 707–35, at 729.
69   De Morgan, review of Brewster, 337.
70   De Morgan's desire for a realistic, or more accurate, approach to biography – which regarded Newton's human foibles and frailties as being no less crucial for a rounded picture of his character than his intellectual accomplishments – has arguably retained its currency to the present day. For example, in her chapter in this volume, Rebekah Higgitt points out some interesting points of historiographical commonality between De Morgan and Newton's most significant recent biographer, Richard Westfall.

Conclusion

It took about a generation after De Morgan's death for his research on the Newton–Leibniz controversy to begin to influence the way in which the affair was portrayed by British (or, more correctly, English-speaking) historians of mathematics. In the first edition of his *Short account of the history of mathematics* (1888), W.W. Rouse Ball still displayed an overwhelming partiality to the claims and conduct of Newton, while disparaging those of Leibniz.[71] But by its fourth edition, he had been forced to add the concessionary sentence: 'During the eighteenth century the prevalent opinion was against Leibnitz, but to-day the majority of writers incline to think it more likely that the inventions were independent.'[72]

One of these writers was Florian Cajori, whose *History of Mathematics* (1894) noted that 'it is generally admitted by nearly all familiar with the matter, that Leibniz really was an independent inventor',[73] even including a supporting quote from De Morgan to reinforce the point. By 1925, David Eugene Smith was able to summarize the calculus controversy in the following words:

> English readers of the 18th century were so filled with the arguments respecting the controversy as set forth in the *Commercium Epistolicum* (1712) and Raphson's *History of Fluxions* (1715), that they gave Leibniz little credit for his work. It was not until De Morgan (1846) reviewed the case that they began generally to recognize that they had not shown their usual spirit of fairness.[74]

---

71 Rouse Ball, *Short Account* (1888), 328–32.
72 Walter William Rouse Ball, *A short account of the history of mathematics* (4th edition, London: Macmillan, 1908), 361.
73 Florian Cajori, *A history of mathematics* (New York: Macmillan, 1894), 233.
74 David Eugene Smith, *History of mathematics* (2 vols, Boston: Ginn and Co., 1923–5), vol. 2, p. 698.

And it is this reading of the dispute that has continued to the present day.[75]

What then were De Morgan's reasons for initiating this revisionist approach to the study of Newton in general and the calculus priority dispute in particular? As we observed at the end of the last section, De Morgan's nonconformist religious beliefs allowed him to divorce considerations of intellectual accomplishments from personal morality. Coupled to this nonconformity was a marked indifference and critical attitude towards such pillars of the British establishment as the Church of England and the Royal Society. Rebekah Higgitt points out that his attitude to the latter 'was just one facet of a stand against the authority of monarchy, aristocracy and established church' and that '[t]hese principles were to inform De Morgan's historical writing, particularly that which related to Isaac Newton – traditionally the hero of establishment science'.[76] According to Higgitt, De Morgan's

> equally passionate belief that religious interests should be kept separate from science and scholarship is demonstrated not only by his decision to teach at UCL – the founding principle of which was that religion should be taught and left at home – but also by his decision to resign his post when he felt that this principle was betrayed[.][77]

It is thus probable that De Morgan's self-imposed exclusion from the British scientific establishment led to a readiness on his part to regard himself as a champion of the underdog – which, in regard to the calculus dispute, Leibniz undoubtedly was in Britain at this time. Another example of De Morgan's advocacy of a similar and contemporaneous *cause célèbre* was his championing of the Italian historian of mathematics Guglielmo Libri, who had been forced to flee France in 1848 on the charge of stealing vast quanti-

---

75 Other more recent studies that give credit to De Morgan for the rehabilitation of Leibniz include, for example, Ivor Grattan-Guinness, 'The British Isles', in Joseph W. Dauben and Christoph J. Scriba, eds, *Writing the history of mathematics: Its historical development* (Basel: Birkhäuser, 2002), 161–78, at 168, and Guicciardini, 'Gigantic implements of war', 729.
76 Higgitt, 'Why I don't FRS my tail', 256.
77 Ibid. See note 31, as well as S.E. De Morgan, *Memoir*, 337–45.

ties of valuable books and manuscripts from French libraries. De Morgan became Libri's most vigorous supporter, publishing a barrage of articles in the British press, wherein Libri was portrayed as a patriot whose belief in Italian precedence had aroused the animosity of the French scientific establishment. But, in this case, De Morgan's unwavering support was not enough to vindicate Libri of the charges against him, primarily because, unknown to De Morgan, Libri was actually guilty as charged.[78]

Perhaps equally significantly, at precisely the same time De Morgan was trying to vindicate Leibniz from unwarranted charges of plagiarism, he too found himself obliged to refute similarly baseless charges relating to his own research. The catalyst was a paper that De Morgan had published in 1846 on the subject of logic, in which he introduced the novel idea of 'numerically definite syllogisms'. This prompted the Scottish philosopher and logician Sir William Hamilton to mistakenly accuse De Morgan of plagiarizing his (quite different) notions on quantifying the predicate, on which he had been lecturing in Edinburgh since the 1830s.[79] The long and drawn-out controversy that ensued with Hamilton and his followers must, in De Morgan's mind at least, have had resonances in the Newton–Leibniz affair. Moreover his own innocence of the charges together with his innate sense of fair play can only have intensified his sense of solidarity with Leibniz and a determination to see justice done.

De Morgan's religious nonconformity thus led not only to his lifelong association with the secular and non-establishment UCL, but to his regarding himself as an outsider in the British scientific community. This had the effect of instilling in him a suspicion of mainstream scientific outlets and academic bodies, as well as a desire to stand up for those he seems to have regarded as fellow outsiders, unjustly accused, or both. A by-product of this antagonistic attitude towards received authority was his relentless research into the Newton–Leibniz dispute, via which De Morgan became the first

---

78   P. Alessandra Maccioni Ruju and Marco Mostert, *The life and times of Guglielmo Libri: scientist, patriot, scholar, journalist, and thief* (Hilversum: Verloren, 1995).
79   Luis M. Laita, 'Influences on Boole's logic: the controversy between William Hamilton and Augustus De Morgan', *Annals of science* 36 (1979), 45–65.

British author not only to dwell on the weak side of Newton's character, but also to attempt to prove that Leibniz had been wronged by the British over the invention of the calculus. In this way, De Morgan's work fostered a sea-change in British attitudes towards the calculus priority dispute, and marked the beginning of a new era in the perception of both Newton and Leibniz by British historians of mathematics.

HENRIK KRAGH SØRENSEN
CENTRE FOR SCIENCE STUDIES, DEPARTMENT OF PHYSICS
AND ASTRONOMY, UNIVERSITY OF AARHUS, DENMARK

Reading Mittag-Leffler's biography of Abel as an act of mathematical self-fashioning

Introduction

In 1903, Gösta Mittag-Leffler (1846–1927) published a biographical essay on the Norwegian mathematician Niels Henrik Abel (1802–1829) in the Swedish magazine *Ord och bild*.[1] Himself a creative Swedish mathematician and an entrepreneur of the discipline, Mittag-Leffler described and commented upon the life of the most illustrious Scandinavian mathematician of the nineteenth century. Mittag-Leffler's essay followed the traditional compositional structure of a biography, thus detailing the events of Abel's life from his birth in 1802 to his death a little more than 26 years later. However, Mittag-Leffler also supplied observations and evaluations from his privileged position as a productive mathematician, a highly regarded journal editor, and an educator keenly interested in nurturing young mathematical talent.[2] The aim of the present paper is to analyze the authorial

---

1   Gösta Mittag-Leffler, 'Niels Henrik Abel', *Ord och bild: Illustrerad månadsskrift* 12 (1903), 65–85, 129–40. Four years later, the essay was translated into French and published in Paris: Gösta Mittag-Leffler, *Niels Henrik Abel* (Paris: La revue du mois, 1907). In the following, all references and quotations are taken from the original Scandinavian-language editions of biographies, and translated by the author.
2   An extensive biography on Mittag-Leffler has recently been published in English: Arild Stubhaug, trans. Tiina Nunnally, *Gösta Mittag-Leffler. A man of conviction* (Berlin and London: Springer, 2010), based on the Norwegian original, Arild

voice in Mittag-Leffler's biography and to contextualize it within Mittag-Leffler's own life, persona, and contemporary context. Thereby, Mittag-Leffler's biography of Abel is not seen as a secondary source for Abel's life, work and times, but rather as a primary source illuminating Mittag-Leffler's efforts at self-fashioning as a legitimate heir to Abel's legacy.

The treatment afforded Abel in biographies of the nineteenth century reflects general trends in the development of the biographical genre and the sub-genre of mathematical biographies. The hagiographic style originally derived from the biographies of saints had parallels in the obituaries of mathematicians when they began to be published by learned societies. Not only did the obituaries often adopt uncritical hero-worship, they also often focused more on the actions and contributions than on the life of the protagonist. This changed with the movement towards the works-and-letters genre in the nineteenth century, aiming to understand the personality of the historical character under study. And with the advent of Freudian psychology, the biographical genre also adopted a deeper appreciation for understanding the formation of the person and his/her character.[3] All

---

Stubhaug, *Med viten og vilje: Gösta Mittag-Leffler (1846–1927)* (Oslo: Aschehoug, 2007); see also the work of Laura E. Turner currently in preparation.

3 For an introduction to the genre of biography, its metaphors, and its relations with history, see for example Barbara Caine, *Biography and History* (Basingstoke: Palgrave Macmillan, 2010); Hermione Lee, *Biography. A Very Short Introduction* (Oxford: Oxford University Press, 2009); Catherine Neal Parke, *Biography. Writing Lives* (New York and London: Routledge, 2002). For some reflections on the genre of scientific biography and its historical development, see the essays of, in particular, David Aubin and Charlotte Bigg, Signe Lindskov Hansen, and Thomas Söderqvist in Thomas Söderqvist, ed., *The History and Poetics of Scientific Biography* (Aldershot and Burlington, VT: Ashgate, 2007); Michael Shortland and Richard Yeo, 'Introduction', in Michael Shortland and Richard Yeo, eds, *Telling Lives in Science: Essays on Scientific Biography* (Cambridge: Cambridge University Press, 1996), 1–44; and Paul Theerman, 'Unaccustomed role: The scientist as historical biographer – two nineteenth-century portrayals of Newton', *Biography* 8 (1985), 145–62. A critical examination and discussion of the genre of mathematical biography is, however, beyond the scope of the present paper.

these features can be found reflected – although sometimes belatedly – in the biographies of mathematicians.

Although mathematicians have long held the history of their subject to be important and integral to further research, their conception of the relevant history had been relatively little concerned with the life and character of great mathematicians beyond the anecdotal. Some of the mathematical journals founded in the nineteenth century also included biographical notices, but these were often short introductions to mathematicians' origins and work derived from national biographies or encyclopaedia.[4] Thus, in mathematics, the biographical and historical genres developed intertwined with both the general cultural and intellectual developments in literary and political biography and history, and with the development of the mathematical discipline itself. This is clearly shown by Rebekah Higgitt in the present volume, where she forcefully argues that the image of Isaac Newton (1642–1727) presented by his biographers developed together with the images of the mathematical discipline and the personal and professional interests of the biographers.[5] It can also be seen from the importance attached to categories such as nation, childhood, ancestry, creativity, and genius in the biographies.

The category of 'genius' is perhaps the most intriguing one among these as it is frequently applied to mathematicians and carries changing

---

4   One such journal to feature biographies was the *American Mathematical Monthly*, started in 1894, which I will address in a separate article. In one of the first periodicals devoted purely to the history of mathematics, the *Bibliotheca mathematica*, the editor Eneström included a few biographical essays, but these were mainly devoted to the biographical facts and the mathematical contributions of mathematicians of the past. Eneström was a fellow Swede and a friend and collaborator of Mittag-Leffler, and his views on the biographical genre will be further studied in another separate paper.

5   For more on the construction of the Newtonian image, see also Niccolò Guicciardini, '"Gigantic implements of war": images of Newton as a mathematician', in Eleanor Robson and Jacqueline Stedall, eds, *The Oxford handbook of the history of mathematics* (Oxford: Oxford University Press, 2009), 707–35; and Rebekah Higgitt, *Recreating Newton: Newtonian biography and the making of nineteenth-century history of science* (London: Pickering & Chatto, 2007).

connotations. In the nineteenth century, the trope of the mathematical genius was often connected to the trope of the romantic hero.[6] In particular, this is the case with the French mathematician and contemporary of Abel, Évariste Galois (1811–1832). Galois' short life has been made the subject of multiple biographies drawing upon the tragic tropes of the devoted but neglected and poor mathematical genius who dies young, before his work can be understood and appreciated by colleagues, so that it is left for posterity to vindicate him.[7] As will become clear in what follows, such tropes were also very often applied to Abel's biography, and it is no coincidence – despite his book's many other faults and inaccuracies – that E.T. Bell's (1883–1960) influential biographies of Galois and Abel were titled 'Genius and stupidity' and 'Genius and poverty', respectively.[8]

When Mittag-Leffler wrote his biographical essay, he was in many ways at the apex of his career and an established researcher, professor, and editor.[9] In his youth, he had studied in Uppsala before going on a European tour to Berlin and Paris, which incidentally had also been the destinations of Abel's tour half a century earlier. In Berlin, Mittag-Leffler studied with Leopold Kronecker (1823–1891) and, in particular, Karl Weierstrass (1815–1897) who each in their way were crucially inspired by Abel's research and results – Kronecker in the theory of equations and Weierstrass in the theory of transcendental functions and the foundations of analysis. Weierstrass is notorious for not always publishing his findings but teaching his newest results to large classes of the best mathematical students in the world. Many of those students later became professors themselves, and

---

6   See for example Amir R. Alexander, 'Tragic Mathematics: Romantic Narratives and the Refounding of Mathematics in the Early Nineteenth Century', *Isis* 97/4 (2006), 714–26. In our manuscript 'The Irony of Romantic Mathematics', Laura Søvsø Thomasen and I also suggest that the notion of 'romantic mathematics' extends beyond the purely biographical.
7   Tony Rothman, 'Genius and Biographers: the Fictionalization of Evariste Galois', *American Mathematical Monthly* 89 (1982), 84–106.
8   Eric Temple Bell, *Men of mathematics* (New York: Simon and Schuster, 1937).
9   For more details and context of Mittag-Leffler's own biography, the reader is referred to Stubhaug, *Gösta Mittag-Leffler. A man of conviction*.

vied for Weierstrass's favour and confidence and the position as his intellectual heir; among them Mittag-Leffler was one of the most fervent. When Mittag-Leffler returned to Sweden in 1881 to become the first professor of mathematics at Stockholm's *Högskola*, he implemented a model of teaching modelled on Weierstrass's courses in Berlin, and in 1884 he succeeded in bringing another of Weierstrass's protégés, Sofia Kovalevskaya (1850–1891), to Stockholm. These actions can be seen as Mittag-Leffler's attempts to appropriate a part of the legacy of the great Weierstrass for himself. In the process, the ties between Abel and Weierstrass also became instrumental. Towards the end of Mittag-Leffler's life, he also made use of the biographical genre to this end when he finally published the rudiments of his biographical project on Weierstrass, which he had worked on for decades.[10]

Since 1881, Mittag-Leffler had very succesfully edited the journal *Acta Mathematica* out of Stockholm. The journal was construed as a Scandinavian enterprise, but in Mittag-Leffler's hands it quickly became very international, and even a mediator between national cultures in mathematics. Its foundation was another occasion when Mittag-Leffler created a link between his own persona and the greatest Scandinavian mathematician of all time. The very first issue was prefixed by a portrait of Abel, and it was Mittag-Leffler's ambition and hope that the young French mathematician Henri Poincaré (1854–1912) would help to promote *Acta Mathematica* in the same way that Abel had propelled August Leopold Crelle's (1780–1855) *Journal für die reine und angewandte Mathematik* to almost instant fame and importance in the 1820s.[11]

Mittag-Leffler's biography of Abel should therefore be seen in relation not only to the biographical genre, but also to Mittag-Leffler's mathematical context including his research trajectory, his efforts at institution building, and his international outlook on mathematics. In the following,

---

10 Gösta Mittag-Leffler, 'Die ersten 40 Jahren des Lebens von Weierstraß', *Acta Mathematica* 39 (1923), 1–57; idem, 'Weierstraß et Sonja Kowalewsky', *Acta Mathematica* 39 (1923), 133–98.
11 Mittag-Leffler was explicit about this comparison in a letter written to Poincaré to solicit papers for *Acta Mathematica*; see Philippe Nabonnand, 'The Poincaré–Mittag-Leffler Relationship', *Mathematical Intelligencer* 21/2 (1999), 58–64, at 59ff.

aspects of Mittag-Leffler's biography of Abel will be shown to have aided in constructing legitimacy for Mittag-Leffler's own persona and thus form part of the self-fashioning of an entrepreneurial mathematician.[12]

Facts and interpretations of Abel's life

Abel was born in 1802 and died at the early age of 26 in 1829.[13] During his short life, Abel studied in Christiania (now Oslo) before embarking on a European tour in 1825–1827 that took him to Berlin and Paris. In Berlin, Abel met the editor Crelle, and in his newly founded *Journal* Abel published the majority of his mathematical works. In particular, Crelle published works by Abel which founded the theory of elliptic functions and which Abel produced in a fierce, yet noble, competition with the German mathematician C.G.J. Jacobi (1804–1851). Abel's entire mathematical production comprises one volume in the collected works first edited

---

12 The notion of self-fashioning has been applied successfully in literary history (see for example Elizabeth Mansfield, 'Emilia Dilke: Self-Fashioning and the Nineteenth Century', in Marysa Demoor, ed., *Marketing the Author. Authorial Personae, Narrative Selves and Self-Fashioning, 1880–1930* (London: Palgrave Macmillan, 2004), 19–39), but less so in the history of mathematics. But with the increased importance given to the modernist author, it is an important cultural phenomenon to attend to, also in the history of science and mathematics.

13 The most comprehensive, recent biographies of Abel to appear in English are Øystein Ore's seminal *Niels Henrik Abel. Mathematician Extraordinary* (Minneapolis: University of Minnesota Press, 1957), based on the Norwegian (Øystein Ore, *Niels Henrik Abel. Et geni og hans samtid* (Oslo: Gyldendal Norsk Forlag, 1954)), and Arild Stubhaug's contextual biography (trans. Richard H. Daly), *Niels Henrik Abel and his Times. Called too soon by Flames Afar* (Berlin: Springer, 2000), based on Arild Stubhaug, *Et foranskutt lyn – Niels Henrik Abel og hans tid* (Oslo: Aschehoug, 1996). All these biographies – and more – can be read using the same approach as employed in the present paper; the present author is currently preparing such a meta-biographical study.

by his teacher Bernt Michael Holmboe (1795–1850) ten years after Abel's death and in a second edition by Ludvig Sylow (1832–1918) and Sophus Lie (1842–1899) in 1881.[14] Abel's stay in Paris was less successful, and he never went to Göttingen to meet Carl Friedrich Gauss (1777–1855) as planned. When he returned to Norway, Abel found no permanent position, but by the time he died in April 1829 after three months of bedridden illness, a position in Berlin was finally materializing.

As even this cursory account of Abel's short life would suggest, his intense life and feverish mathematical production are the stuff that feeds legends of genius. Abel suffered financial hardship and insecurity about his professional future and his family situation. At the same time, his new mathematical ideas were often difficult to comprehend for the greatest mathematicians of his time. And yet he continued to found and develop theories that would become very important for the development of mathematics during the nineteenth century: in particular the theory of elliptic functions and their generalization into so-called abelian functions. Many of the key elements for the romantic mythology of Abel as a neglected yet defiant mathematical martyr were established during the first half-century after his death, in obituaries and short biographical notices written by people who had actually known him.

The first work to present a comprehensive biography of Abel was published by the Norwegian mathematician Carl Anton Bjerknes (1825–1903)

---

14 Niels Henrik Abel, ed. Bernt Michael Holmboe, *Oeuvres Complètes de N.H. Abel, mathématicien, avec des notes et développements* (2 vols, Christiania: Chr. Gröndahl, 1839) and Niels Henrik Abel, ed. Ludvig Sylow and Sophus Lie, *Oeuvres Complètes de Niels Henrik Abel* (2 vols, Christiania: Grøndahl, 1881), respectively. For an analysis of Abel's mathematics, see for example Christian Houzel, 'The Work of Niels Henrik Abel', in Olav Arnfinn Laudal and Ragni Piene, eds, *The Legacy of Niels Henrik Abel. The Abel Bicentennial, Oslo, 2002* (Berlin: Springer, 2004), 21–177; and Henrik Kragh Sørensen, *The Mathematics of Niels Henrik Abel: Continuation and New Approaches in Mathematics During the 1820s*, Research Publications on Science Studies 11 (Department of Science Studies, University of Aarhus, 2010), <http://www.ivs.au.dk/reposs>.

in 1880 and in a French edition in 1885.[15] The original Norwegian version spanned 130 pages but was extended to 360 in the French version. A central point of Bjerknes's biography was to investigate Abel's fierce competition with Jacobi, and through meticulous collection and untangling of the chronology of events to lay a major claim for priority on Abel's behalf. Such topoi of rivalry and conflict are, of course, regular and important features in biographies. In the case of Bjerknes's biography, it had been spurred by a recent publication of the German mathematician and biographer Leo Königsberger (1837–1921), which Bjerknes felt encroached on Abel's priority.[16]

Bjerknes's approach to writing Abel's life was well in line with the traditional, empiricist view of the biographical genre still in vogue at the time;[17] he was documenting *what Abel had done*, being slightly less interested in *who Abel was*. This latter more empathetic approach was adopted by the Norwegian mathematician and author Elling Holst (1849–1915). The occasion was the publication of the extant letters from Abel in a *Festskrift* issued in Norwegian and French for the centennial celebrations of Abel's birth in 1902.

According to Holst, the reason for publishing the complete collection of Abel's letters was that, in their entirety, they provide inescapable and essential insights into his personality and the conditions under which he had to live.[18] Thus, Holst's focus was not such much to analyse Abel's mathematics as to paint a portrait of Abel's character and personality.

---

15   Carl Anton Bjerknes, 'Niels Henrik Abel. En skildring af hans liv og videnskabelige virksomhed', published as an appendix to *Nordisk tidsskrift för vetenskap, konst och industri* (Stockholm: P.A. Norstedt & Söner, 1880); idem, *Niels-Henrik Abel. Tableau de sa vie et des on action scientifique* (Paris: Gauthier-Villars, 1885).

16   Leo Koenigsberger, *Zur Geschichte der Theorie der elliptischen Transcendenten in den Jahren 1826–29* (Leipzig: B.G. Teubner, 1879).

17   See for example Thomas Söderqvist, 'Introduction: A New Look at the Genre of Scientific Biography', in Söderqvist, *Scientific Biography*, 1–15, at 5.

18   In Holst's actual words, '*the true* picture': Elling Holst, 'Niels Henrik Abel. Historisk indledning til hans efterladte breve', in Elling Holst, Carl Størmer, and Ludvig Sylow, eds, *Festskrift ved Hundredeaarsjubilæet for Niels Henrik Abels Fødsel* (Kristiania: Jacob Dybwad, 1902), 3. In short, Holst maintained, 'the letters are the necessary supple-

For the *Festskrift*, Holst provided the letters with an extensive introduction spanning 110 pages and amounting to a full biography of Abel.[19] In this introduction, he not only let the letters guide the biography, he also provided information and perspectives of his own. Thus Holst pointed out how Abel had unknowingly drawn himself through the few letters which give a silhouette of his 'inner person', clearly bringing out the contour of the soft, childlike features of his innermost soul.[20]

> In order to understand the letters, his life has to be briefly told. And in order to understand his life, it must be seen against a background of the conditions under which he was born, was formed, and worked.[21]

The condition most important for Holst to describe was the troubled position of Norway in the wake of the Napoleonic wars and the Peace of Kiel (1814), which severed Norway from Denmark. Thus Holst's biography was cast in patriotic terms of direct relevance to the celebratory *Festskrift* and the contemporary political situation in Norway.[22] As Holst put it in the very first sentence of the biography: 'Abel's birth and childhood fell during war times; [when he was 12 years old,] his fatherland almost miraculously

---

ment to concisely illuminate his short life in poverty which stood in such contrast to his unique genius' (ibid.).

19   Ibid.; it was published in the French edition as Elling Holst, 'Niels Henrik Abel. Introduction historique a sa correspondance', in Elling Holst et al., eds, *Niels Henrik Abel. Memorial publié à l'occasion du centenaire de sa naissance* (Kristiania: Jacob Dybwad etc., 1902).

20   Holst, 'Historisk indledning', 3.

21   Ibid., 4. Actually, the original word 'vaagned' means 'awoke', thus clearly drawing upon a myth of an unrecognized genius that needs to be woken up.

22   Towards the end of the nineteenth century, patriotic sentiments were widespread in Norway, as were calls for increased independence or separation from the union with Sweden. The role of the centennial celebrations in 1902 within such a political context is also discussed in Henrik Kragh Sørensen, 'Niels Henrik Abel's Political and Professional Legacy in Norway', in Reinhard Siegmund-Schultze and Henrik Kragh Sørensen, eds, *Perspectives on Scandinavian Science in the Early Twentieth Century* (Oslo: Novus forlag, 2006), 197–219.

became an independent realm'.²³ The Norwegian patriotism in the mid-1810s, as well as the financial and cultural hardship of the young nation, then became recurring themes in Holst's portrayal of the background for understanding Abel's life.²⁴

Thus, when Mittag-Leffler published his biographical essay on Abel in 1903, the basic facts about Abel's life and work were well established through Bjerknes's extensive biography, the full publication of Abel's letters accompanied by Holst's introductory commentary, and the annotated edition of Abel's complete works edited by Sylow and Lie.

The context of Mittag-Leffler's biography

In contrast to the biographies by Bjerknes and Holst, Mittag-Leffler's biographical essay was intended to reach further beyond the mathematically-inclined. At 33 pages including illustrations,²⁵ it was considerably shorter than its two main precursors, and in certain ways an intermediary between those and shorter biographies of Abel published in magazines and literary encyclopaedias.²⁶ Published since 1892, *Ord och bild* was an illustrated magazine issued monthly, bringing together leading Swedish and Scandinavian artists and writers in an independent cultural magazine.²⁷

23 Holst, 'Historisk indledning', 4.
24 Ibid., 3–4.
25 Published without illustrations, the French edition spans 48 pages.
26 The first such biography was written by Christopher Hansteen: 'Niels Henrik Abel', *Illustreret Nyhedsblad* 11/9–10 (2–9 March 1862), 37–8, 41–2. When Holst contributed a chapter on mathematics to the illustrated Norwegian literary encyclopaedia, 10 pages out of the total 34 comprised a biography of Abel: Elling Holst, 'Mathematik', in Henrik Jæger and Otto Anderssen, eds, *Illustreret norsk literaturhistorie. Videnskabernes litteratur i det nittende aarhundrede* (4 vols, Kristiania: Hjalmar Biglers Forlag, 1896), vol. 4, pp. 68–101.
27 Largely independent of financial or partisan political interests, the magazine developed a 'radical' and progressive position among the Swedish intelligensia during the

Despite its differences from the two biographies by Bjerknes and Holst, it is in relation to them that Mittag-Leffler's essay must be understood and analyzed for its argumentative or programmatic function.[28] This can be done by focusing on the authorial voice as it is manifested in the text through deliberate choice of material and narrative structure, style, rhetorical figures, pronouns, and forms of language.[29]

As the reviewer and editor Emil Lampe (1840–1918) noticed in his review of the French edition for the *Jahrbuch über die Fortschritte der Mathematik*, Mittag-Leffler's biography did not go beyond the works of Bjerknes and Holst.[30] Yet he did point to one novel piece of information, namely that it was at Mittag-Leffler's instigation that Bjerknes had undertaken his efforts to sort out Abel's priority over Jacobi, which eventually led to his biography. This episode can be seen as indicative of the way Mittag-Leffler used his biographical writing to position himself at the heart of knowledge of, and appreciation of, Abel. During his student days in Uppsala, Mittag-Leffler was frustrated to learn of Abel's greatness only indirectly and not through the mathematics teaching, which he found very elementary.[31] When Mittag-Leffler subsequently began work on the

> 1890s. In 1903, when Mittag-Leffler's biography was published across two separate issues, the magazine also featured pieces of artwork, poems, and essays on such a diversity of issues as the museum of natural history in London, the new trend of public libraries, and Swedish poetry and prose.
>
> 28 For another case study in the argumentative functions of biographies, see Signe Lindskov Hansen, 'The Programmatic Function of Biography: Readings of Nineteenth- and Twentieth-Century Biographies of Niels Stensen (Steno)', in Söderqvist, *Scientific Biography*, 135–53. Together with a study on Abel's genealogy, the works of Bjerknes and Holst are the only references explicitly given by Mittag-Leffler.
>
> 29 See for instance David Aubin and Charlotte Bigg, 'Neither Genius nor Context Incarnate: Norman Lockyer, Jules Janssen and the Astrophysical Self', in Söderqvist, *Scientific Biography*, 51–70.
>
> 30 Emil Lampe, review of Mittag-Leffler, *Niels Henrik Abel*, in *Jahrbuch über die Fortschritte der Mathematik*, 40.0020.01 (1908).
>
> 31 This inspired Mittag-Leffler to write a broadly accessible biography of Abel to make his life and work more known at home, and he started the project during the 1870s. In 1874, when Sylow and Mittag-Leffler met in Berlin, Mittag-Leffler spoke

definition and history of elliptic functions and travelled to Germany, he had become interested in the events by which Abel and Jacobi had founded the theory. And when in 1875 he wrote to ask Bjerknes for help in accessing one of Abel's manuscripts, he became aware that Bjerknes was preparing a biography of Abel to be published in instalments in a newspaper. Spurred by Mittag-Leffler's interest – so Mittag-Leffler lets us believe – Bjerknes extended and elaborated his studies into the full biography published five years later.[32] Mittag-Leffler had also been interested in writing a biography of Abel back then, but upon learning of Bjerknes' project he gave up on the idea and left it to him, who 'had completely different personal and even national prerequisites' for writing the biography. Thus, according to Mittag-Leffler himself, nationality was something that set him apart from the perspective of Bjerknes (and Holst).[33] As an outsider in national terms,[34] Mittag-Leffler would present a different, much more non-national perspective on some of the important and controversial points in Abel's biography.

Mittag-Leffler and patriotic biography

In multiple ways, Mittag-Leffler's biography of Abel responded to the previous ones by addressing some of their central topics while providing them with a slightly different spin. This is the case with Bjerknes' concern over priority and with Holst's concern over the national character and patriotic importance of Abel's work. Both of these topics share a dual embedding in

---

enthusiastically about his plans for a biography of Abel: Stubhaug, *Med viten og vilje*, 203.
32   Mittag-Leffler, 'Niels Henrik Abel' (1903), 132.
33   Ibid.
34   Although the union between Sweden and Norway still existed, Norway had considerable independence and a growing national awareness.

the national and patriotic contexts in Norway during Abel's times as well as during the times of the biographers.

Priority and the role of an editor

A considerable part of Bjerknes' biography of Abel and a large part of its reception in European mathematical circles centred around the fierce competition with Jacobi and the issue of priority for the discovery of elliptic and higher transcendental functions. At a time when Norwegian intellectuals, scientists, artists, and explorers were seeking opportunities to claim a role for Norway as a cultural nation, securing Abel's priority in mathematics was almost a national matter.[35]

35 During the last decades of the nineteenth century, artists and poets such as Munch, Ibsen, Grieg and Bjørnson were becoming increasingly known abroad while being clearly embedded in the Norwegian culture of their times. Historians such as Sars began researching and presenting Norwegian culture and history, and a new language (*nynorsk*) was devised and given equal status with the old colonial Danish (*bokmål*) in 1885. But nowhere was the intellectual and cultural nation-building more prominent than in the expeditions to the Arctic led by Nansen in 1893 and Amundsen after the turn of the century (for the role of science in Norwegian nation-building, see also Geir Hestmark, *Vitenskap og nasjon. Waldemar Christopher Brøgger 1851–1905* (Oslo: H. Aschehoug & Co., 1999), and Rune Slagstad, *De nasjonale strateger* (Oslo: Pax Forlag, 1998)). The national importance is clearly visible in Bjørnson's very popular commemoration (1896) of Nansen's deed, which – tellingly, perhaps – in the national edition of Bjørnson's works is printed directly before the cantata from the Abel-centennial (see below). There, Bjørnson's poetic talents were employed to the good of Norwegian national understanding, mentioning 'Norway' in three of the four stanzas: Bjørnstjerne Bjørnson, *Samlede værker. Mindeutgave* (5 vols, Kristiania and Kjøbenhavn: Gyldendalske Boghandel & Nordisk Forlag, 1910–1911), vol. 1, p. 170). The great national stakes involved are also clearly discernible from the reviews of Bjerknes' book in German and French mathematical journals; see for instance Joseph Bertrand, review of Bjerknes, *Niels-Henrik Abel*, in *Bulletin des sciences mathématiques* (second series) 9 (1885), 190–202; G. Brunel, review of Bjerknes, *Niels-Henrik Abel*, in *Bulletin des sciences mathématiques* (second series) 9 (1885), 141–53; Eugen Netto, review of Bjerknes, *Niels-Henrik Abel*, in *Jahrbuch über die Fortschritte der Mathematik* 17.0014.04 (1885).

Publishing in 1879, Königsberger based his detailed study of the genesis of the theory of elliptic functions and their generalizations on the published works by Jacobi and Abel, as well as published correspondence and published material from the *Nachlass* of Gauss. From these sources, Königsberger concluded that the birth of the theory of elliptic functions was due to two independent and profound discoveries reached by the two great young mathematicians, and had partially been anticipated by Gauss.[36] However, the shared claim to fame became a serious provocation for Bjerknes to rebuke, by carefully re-analyzing the chronology of events and drawing inferences and interpretations from it.[37]

On this issue of priority and pride, Mittag-Leffler adopted a surprisingly different perspective, referring to Gauss's assessment of Abel's work as 'astounding words to all those small professors who dabble over priority, in their futile race for the unattainable ideal'.[38] Thus Mittag-Leffler turned against Bjerknes's attack on Jacobi and his followers among German mathematicians for neglecting the role of Abel in the early years of the theory of elliptic functions. After all, Jacobi lived much longer and published a treatise, *Fundamenta nova*, which founded the symbolic framework for dealing with elliptic functions in the very year that Abel died.

Instead of disputing the series of events that led to the formulation of the theory of elliptic functions, Mittag-Leffler focused on the differences in style between Abel and Jacobi:

---

36 See for example Koenigsberger, *Zur Geschichte der Theorie der elliptischen Transcendenten*, 43–4, 104. Koenigsberger's study had been prompted by the fiftieth anniversary of the publication of Jacobi's seminal work, the *Fundamenta nova*, which appeared at the same time as 'the death of Abel, the other great creator of the theory of transcendentals' (ibid., preface).

37 This aspect of Bjerknes' biography is, obviously, also addressed at more length in the study under preparation. For Holst's representation of the competition with Jacobi, see Holst, 'Historisk indledning', 88–93.

38 Mittag-Leffler, 'Niels Henrik Abel' (1903), 132.

Jacobi was a great mathematician with a brilliant formal intellect and an unsurpassed mastery of the language of mathematical formulae, but he was inferior to Abel in ingenuity and power of thought.[39]

Mittag-Leffler clearly valued the genius of his protagonist over the hard-working style of Jacobi; and to him Jacobi and Abel represented different, complementary approaches to the study of elliptic functions. Whereas Jacobi's approach had been the most developed one in the middle of the nineteenth century, Weierstrass's work on the theory and his general approach had highlighted the merits of Abel's point of view as well.

Concerning the frantic development of results and ideas during the fall of 1827, Mittag-Leffler also positioned himself differently from Bjerknes. Drawing upon his own long experience as editor of *Acta Mathematica*, Mittag-Leffler added his perspective on the relations between Crelle and Jacobi. It was known that in connection with founding his *Journal* in 1826 Crelle had attempted to solicit contributions from Jacobi and they had corresponded extensively. Thus, Bjerknes had been led to suspect that Crelle had informed Jacobi about Abel's elegant and very fertile new approaches to the study of elliptic integrals. If so, this would be the source for the dependence of Jacobi's earliest contributions upon the key ideas in Abel's papers.

However, far from meddling with the petty issues of priority, Mittag-Leffler adopted a different perspective: 'Nothing would have been more natural'.[40] Thus what to Bjerknes might have appeared as a blatant indiscretion on the part of Crelle was completely natural to Mittag-Leffler. And, based on his experience and reputation as the editor of Scandinavia's foremost mathematical journal, Mittag-Leffler's view would in 1903 have carried some weight. Thus, the importance of journals – and their editors – for Mittag-Leffler would seem to lie not in the securing of priority, but rather in the development of scientific thought; national or individual interests were thus subsumed under the greater, non-national progress of science. Where Bjerknes had gallantly sought to defend the national hero against

39 Ibid., 133.
40 Ibid., 131.

encroachment on his priority, Mittag-Leffler pointed out the development of science as the main interest, thereby relegating petty disputes over priority to the level of inferior mathematicians. Thus, Mittag-Leffler, who had seen Bjerknes as 'uniquely qualified from a national perspective' to write the biography of Abel, could adopt the perspective of a partial outsider whose interests were more lofty and trans-national than the fight for priority and national pride.[41]

The immature Norwegian state and an outsider's perspective

A separate topic, of great importance even in the first obituaries of Abel written by his friends, was the role played – or not played – by the Norwegian state in alleviating Abel's financial situation and securing the great mathematician for the young nation. The Norwegian government had supported Abel's tour abroad, but when one of the only two professorships in mathematics in Christiania became vacant while Abel was away, Abel's teacher and friend Holmboe was appointed. Thus, upon his return, Abel found no openings for an academic career in Norway, and he had to make do with minor and irregular teaching jobs. In his addition to Holmboe's obituary of Abel published in a Norwegian scientific magazine, Abel's friend and travel companion Christian Boeck (1798–1877) lamented this passiveness and narrow-mindedness of the Norwegian authorities, both academic and political, in securing the great genius for the nation and – between the lines – blamed this neglect for Abel's untimely death.[42] This was the birth of an enduring idea: that Abel's poverty was so great it eventually

---

41 As is known from his work with securing his priority and that of his students, Mittag-Leffler was not consistently so altruistic in his views about the cultivation of science, as Laura E. Turner shows in 'A Man on a Mission: Mittag-Leffler's Efforts to Promote the Research Imperative to His Students at Stockholm's Högskola' (unpublished draft, May 2010).

42 Footnote by Boeck inserted in Bernt Michael Holmboe, 'Necrolog. Kort Fremstilling af Niels Henrik Abels Liv og videnskabelige Virksomhed', *Magazin for Naturvidenskaberne* (1829), 334–54, at 340–1. By 1829, Boeck was the editor of the magazine.

killed him, and that the Norwegian authorities, through their passiveness, hastened the process.[43]

Holst interpreted the appointment of Holmboe as an untimely act of conservatism on the part of the Norwegian academic elite. To Holst, the academic elite sought to promote a talented man whose intellect they could understand, rather than the genius abroad.[44] Thus, Holst points directly to his view of the mediocre state of Norwegian society at a time when 'We, as a society, had simply not progressed any further'.[45] For Holst, Abel had been seriously and undeservedly *wronged* by this decision to prioritize domestic teaching obligations over the support of his excellent research.[46]

One particular tragic event had fostered criticism of the government for finding no appropriate position for Abel, yet giving the one position which was deemed appropriate for him to Holmboe. Just two days after Abel had died in Froland, and before the news had reached Berlin, his friend and editor Crelle wrote to say that a position had finally been secured for him Berlin.[47] This followed a prolonged effort to call Abel to Berlin, that had taken place while he was struggling in Norway. Since the international experts could apparently see and appreciate Abel's qualities, the inaction of the Norwegian authorities seemed to the patriotic Boeck and Holst provincial and narrow-minded.

In his biography, Mittag-Leffler also commented upon the issue of national support for Abel, from a different perspective. He saw the travel grant issued by the Norwegian government in 1825 as a remarkable piece of support on the part of political institutions for an individual scientist and

---

43   Abel had probably contracted tuberculosis during his European tour, and although he complained about the lack of money, the illness that directly caused his death was probably the result of a winter excursion through Norway to visit his fiancée for Christmas.
44   Holst, 'Historisk indledning', 43.
45   Ibid., 44.
46   Ibid., 109–10.
47   Crelle's notorious letter to Abel sent from Berlin, 8 April 1829 (Niels Henrik Abel *et al.*, 'Breve fra og til Abel', in Holst, Størmer and Sylow, *Festskrift*, at 89–90) is frequently used by biographers to construct the climax of a tragic narrative; see for example Holst, 'Historisk indledning', 105.

for science in general: 'There are, in the history of the Nordic countries, few acts of government which have had a greater impact for science'.[48] Similarly, his evaluation of the actions of the Norwegian authorities during Abel's last summer in Norway differed from Holst's.

> The academic authorities cannot be blamed. They did what they could. And anyone who is used to the slow procedures of such matters in Sweden must even admire the speed with which one communiqué followed another.[49]

Thus, Mittag-Leffler not only defended the Norwegian authorities on this matter, but also used the opportunity to comment upon the slow and cumbersome procedures that he suffered, himself, in his own daily dealings with the Swedish bureaucracy.

Before the call to Berlin finally came, Crelle and Abel had been frustrated in an earlier set-back which had taken place when Abel had already started informing his friends and mentors of success. Such a public disclosure and ensuing polemic about delicate and confidential career choices were astonishing to Mittag-Leffler, as was the public defence of the university in Christiania against its alleged neglect of Abel, published in a leading Norwegian newspaper in 1829. After outlining the events and efforts on Abel's part and their eventual failure, Mittag-Leffler could not resist the following more personal remark:

> [...] this entire phase in Abel's history, with the public debate of cases of the most confidential nature, strongly resembles similar recent events in our brother-land, which always provoke great astonishment here in Sweden.[50]

Mittag-Leffler's comparison was also aimed at very public debates about the appointment of academics during the last decades of the nineteenth century, when the Norwegian cultural and academic elites mutually bolstered their influence through publicly-debated and politically-motivated

---

48 Mittag-Leffler, 'Niels Henrik Abel' (1903), 74.
49 Ibid., 130–31.
50 Ibid., 135–6.

appointments.[51] Thus, the claim of provincialism was turned on its head by Mittag-Leffler: it was not the failures of the local administration in Oslo, but rather the overly open public debate about personal and controversial matters, that made Norway provincial in his mind. Thus, it was not the government but rather its critics, including Boeck and even Abel himself, who did not know how to play the political game well.

On one particular point, though, Mittag-Leffler had to strongly disapprove: the decision to appoint Holmboe as replacement for professor Søren Rasmussen (1768–1850). The main obligation of the position was the teaching of elementary mathematics to first-year students of the University. Although Holmboe was suggested for the position from the very beginning, attention was also drawn to Abel by his mentors among the professors. However, two circumstances were advanced against Abel as a candidate. First, he was at the time still abroad on his European tour, and it would be a considerable loss to the progress of his studies if he were to be called back to Norway before the end of the tour. And second, it was assumed that he would not find it as easy to accommodate the intellectual powers of the young students as a more experienced teacher would.

The first point could be seen as a positive one, shielding the young and hard-working mathematician and perhaps implicitly suggesting that he could be found a job upon his eventual return to Norway.[52] This was not remarked by Mittag-Leffler, who took serious exception to the second part of the argument.

51 The first such politically-ordained position was given to Lie to preserve him in Norway when he had the possibility of going abroad; see Arild Stubhaug, trans. Richard H. Daly, *The Mathematician Sophus Lie. It was the Audacity of My Thinking* (Berlin: Springer, 2002), 153ff. In that connection, Abel's legacy was directly used as a rhetorical figure; see Sørensen, 'Niels Henrik Abel's Political and Professional Legacy'. Normally, of course, the appointment of professors were the privilege of the government, and not of Parliament.
52 Such seems to have been the ambition of Abel's mentor, Hansteen, who was involved with writing the recommendation.

> The reasoning of the faculty [in Oslo] is just as common as it is wrong. It is assumed that a mere mediocrity can better adjust to the 'capabilities of comprehension of the younger students' than the truly clever[.][53]

Abel's writings were, for Mittag-Leffler, unique in their clarity and elegance, and this seemed to him to promise a clear style in teaching as well. Indeed, Abel might have been particularly suited to teach the young and unformed, he argued. Abel's troubles in communicating his groundbreaking new ideas were always with the older generation, not with the young.[54] The education of mathematicians had been a crucial point for Mittag-Leffler ever since he was appointed the first professor of mathematics at Stockholm's new *Högskola* in 1881. There he developed an instructional system partly modelled on Weierstrass's lectures on analysis in Berlin, aimed at training and engaging the brightest students in mathematical research.[55] With this background, he advanced new standards and aims for teaching which also incorporated severe criticism of traditional teaching in mathematics.

> Every true mathematician knows how much more difficult it is to come to terms with older students who have already had a mediocre or poor education than with the younger ones whose intellects have not yet been clouded by obscure knowledge.[56]

Mittag-Leffler connected this with a similar point made by Weierstrass in a letter to Kovalevskaya: Weierstrass had long given up the hope of influencing his older colleagues; it was the young to whom he addressed himself.[57] This emphasis on training the young generation of mathematicians was taken up as a credo by Mittag-Leffler himself.[58]

---

53 Mittag-Leffler, 'Niels Henrik Abel' (1903), 78.
54 Ibid.
55 Discussed in Turner, 'A Man on a Mission'.
56 Mittag-Leffler, 'Niels Henrik Abel' (1903), 79.
57 Ibid.
58 Discussed in Turner, 'A Man on a Mission'.

National pride and patriotic obligation

Whereas Bjerknes's biography had primarily comprised empiricist documentation of Abel's actions and the events of his life, Holst's wanted also to convey a picture of Abel's psyche – of the 'innermost soul' of the Norwegian mathematician. The serious and existential set-back caused by the decision to appoint Holmboe – 'the cross-road in Abel's life', as Holst described it – fed directly into Holst's reading of Abel's psyche.[59] Abel travelled together with a small group of young and promising Norwegian scientists on their great educational tour, and Holst paints a picture of Abel highlighting his desire to be surrounded by fellow Norwegians. Drawing on Bjerknes, Holst described how Abel's mood had at first been joyous in the Norwegian camp of young travellers, but he would later become passive, silent and depressed for days, describing himself as 'in a dark mood'.[60] Rhetorically, in Holst's biography, the change was brought about by the dire news from home.[61] Abel's depressive mood only gained in strength as he later found himself alone in Paris without the group of Norwegian comrades around him.[62]

As Holst found, there are relatively few expressions of aesthetics and culture in Abel's letters,[63] but he did seize upon value-laden descriptions of Norway and its natural landscape. At times, Holst strains to interpret Abel's yearnings for the Norwegian landscape as expressions of patriotism, as for example where he a little uncritically interprets Abel's comparison of the breathtaking view of the Adriatic Sea to the view at home from Ekeberg.[64] Writing to Holmboe, who had never travelled abroad, Abel described how the small group had approached the Adriatic Sea and Trieste and discovered a view that was 'indisputably very beautiful, yet by no means comparable to that from Ekeberg'.[65] Thus, it could also be that

---

59   Holst, 'Historisk indledning', 43.
60   Ibid., 46.
61   Ibid., 48, in particular.
62   Ibid., 63.
63   Ibid., 52–6, in particular.
64   Ibid., 51.
65   Abel to Holmboe, Bolzano, 15 June 1826: Abel, Niels Henrik *et al.*, 'Breve fra og til Abel', 33–7, at 34.

Abel's mention of Ekeberg or his comparison of Steiermark around Graz in Austria to the Norwegian landscape were not so much expressions of his patriotism and delight in the Norwegian landscape as simple and useful comparisons to help people at home to comprehend the foreign spectacles the young Norwegians were encountering.[66] There are also comparisons of less spectacular matters, such as the 'sterile landscape' in Mähren or the German mail service, to Norwegian counterparts.[67] Holst's biographical programme of bringing the inner Abel out through his letters ran the risk of distortion through inattention to their contexts and through too directly – and without considering the genre – drawing implications about Abel from his travel reports.

Mittag-Leffler reacted to some of Holst's empathetic and psychologically sensitive interpretations. Where Holst had seen a budding patriotic (if not outright nationalistic) awareness in Abel, he sought other explanations. This is clearest in respect of Holst's patriotically-imbued interpretation of Abel's decision to return home and seek a position in Norway, even after Holmboe had been appointed to the only likely job that would suit him. In his biography in the *Festskrift*, tied to the national celebrations, Holst had explained Abel's decision in terms of a sense of duty to the nation: Abel, who was rewarded with a state stipend for his European tour, simply could not conceive of the possibility of not returning; it was an unquestionable patriotic obligation for him.[68] Mittag-Leffler saw this as a specifically Norwegian, domestic issue, and found it grossly exaggerated. He saw Abel's desire to return to Norway in terms of his temperament and his difficulty in coping with strangers.[69]

Mittag-Leffler thus repeatedly used his dual position as an outsider to the national and patriotic polemics, and as an insider to the mathemati-

---

66   The comparison with Steiermark is made in the same letter; see p. 33.
67   Abel to Holmboe, Wien, 16 April 1826, in ibid., 26–31, at 29, and Abel to Holmboe, Bolzano, 15 June 1826, in ibid., 33–7, at 33.
68   Holst, 'Historisk indledning', 70.
69   Ibid., 85. As indicated above, these character traits had also been noticed by Holst but had been played down in favour of patriotism as explanations for Abel's decision to return.

cal enterprise of research and publishing, in interpreting and commenting upon Abel's biography. Mittag-Leffler's biographical essay must therefore be seen as a commentary on the earlier biographies, made possible by this new combination of perspectives. It is telling that whereas many Norwegian biographies of Abel have quoted the final stanza of Bjørnstjerne Bjørnson's (1832–1910) poem composed for the centennial celebrations, Mittag-Leffler prefaced his with an excerpt from a different stanza. Instead of the highly patriotic 'Now he is owned by the world / but the boy was ours', he chose this: 'Where he set foot / cannot be the same again', repeating his preference for the topos of genius over that of patriot.[70]

Mittag-Leffler as heir to Abel's legacy

Mittag-Leffler's assessment of Abel's mathematics

Mittag-Leffler's presentation of Abel's mathematical contributions merits special attention because of the particular 'spin' and emphasis that he applied. In assessing the extent and scope of Abel's work, he wrote:

> Abel's works comprise no more than a large quarto volume. In its size, Abel's *Oeuvres* is well below that of other great mathematicians. But what a massive world of new thoughts that are contained in this volume. There is hardly any mathematical work of importance conceived after Abel which is not to a smaller or larger extent influenced by him. The biggest contributions of the past century – the theory of analytic functions and the theory of abelian functions – are direct and immediate continuations of his work.[71]

---

70 'Der han var / kan tenkes ikke uden ham': ibid., 65. Bjørnson's poem was published in Norwegian and French in Holst, Størmer, and Sylow, *Festskrift*; see Elling Holst *et al.*, eds, *Niels Henrik Abel. Memorial publié à l'occasion du centenaire de sa naissance* (Kristiania: Jacob Dybwad etc., 1902); the English translation of the last stanza is taken from Stubhaug, *Niels Henrik Abel and his Times*, 14.
71 Mittag-Leffler, *Niels Henrik Abel* (1903), 139.

Thus Mittag-Leffler praised – as would many others – the depth over the breadth of Abel's work. And that notion of depth would apparently be understood in terms of 'ideas' and 'influence'. Thus, the deep thoughts that Abel had presented were contributions to science, and their anchoring in the existing mathematical literature was not very much emphasized. Instead of seeing Abel as an end- or midpoint of some development, Mittag-Leffler's biography presented him as an originator of ideas to be developed by later mathematicians. In this way, the persona of the groundbreaking mathematical genius is recurrent and prominent in Mittag-Leffler's biography of Abel.

In the quotation above, Mittag-Leffler singled out the fields of analytic and abelian functions as the biggest contributions of the nineteenth century, and therefore also as the most important foundation stones set down by Abel. But, as Mittag-Leffler also briefly outlined, Abel's mathematical production spanned the field of elliptic functions, Abelian integrals, the theory of algebraic solvability of equations, the rigourisation of analysis and other branches of mathematics.[72] Among these, Mittag-Leffler singled out just two fields, which were particularly important to him and his own research agenda.

Abel's work in the field of analytic and abelian functions involved his theory of elliptic functions as inverse functions of elliptic integrals, as set forth in his seminal papers entitled *Recherches sur les fonctions elliptiques*, of 1827 and 1828.[73] This theory, the subject of the competition with Jacobi and the subsequent question of priority, was described and discussed by Mittag-Leffler from a different perspective. First, he praised Abel's theory of elliptic functions as a 'rounded theory built from scratch and aimed at completeness', something that quite set it apart from Jacobi's more haphazard deduction of particular results in the theory of transformations.[74]

---

72 See Sørensen, *The Mathematics of Niels Henrik Abel*.
73 Niels Henrik Abel, 'Recherches sur les fonctions elliptiques', *Journal für die reine und angewandte Mathematik* 2/2 (1827), 101–81; idem, 'Recherches sur les fonctions elliptiques', *Journal für die reine und angewandte Mathematik* 3/2 (1828), 160–90 (both reproduced in Abel, *Oeuvres Complètes* (1881), vol. 1, pp. 263–388).
74 Mittag-Leffler, *Niels Henrik Abel* (1903), 131.

The theoretical basis and completeness of Abel's theory were attributes that Mittag-Leffler valued highly around 1900, when these characteristics resonated with ideas about 'modern' mathematics (see further below).

But the characterization of Abel's theory as rounded and complete allowed another impressive argument as well. With reference to the way the elliptic functions were defined and introduced at the outset of the *Recherches*, Mittag-Leffler explained how 'in the proof there is – at a certain point – a lacuna which I have elsewhere shown how to fill without difficulty and using Abel's own line of thought.'[75] The reference points to Mittag-Leffler's own first really independent piece of mathematical research, which he developed for his qualifying lecture for a professorship competition in Helsinki and published in Swedish as *A method of obtaining the elliptic functions* in 1876.[76] The topic had apparently been suggested to him by Weierstrass during their discussions in Berlin; in the paper Mittag-Leffler devised a different way of defining Abel's elliptic functions.[77] This deserved mentioning in his 1903 biography of Abel because Mittag-Leffler thereby not only added to the work of the great innovator but actively perfected Abel's 'rounded and complete' theory of elliptic functions. Simultaneously, he consolidated an intellectual genealogy from Abel through Weierstrass to himself.

Abel as a modern mathematician

To Mittag-Leffler, Abel was not only an extremely productive and innovative mathematician, setting out and developing new theories and fields. For someone who was himself keen on the modernizing of Scandinavian mathematics around 1900, Abel emerged as the herald of new conceptions about mathematics and its practice.

---

75 Ibid.
76 Gösta Mittag-Leffler, *En metod at komma i analytisk besittning af de Elliptiska Funktionerna* (Helsingfors: J.C. Frenckell & Son, 1876).
77 Stubhaug, *Gösta Mittag-Leffler. A man of conviction*, 212.

By tackling profound questions in his chosen fields, and by critically reflecting on the new approaches of thinkers like Augustin-Louis Cauchy (1789–1857), Abel had 'looked too far into the innermost connections to not recognize that even his intuition needed the control of rigorous deduction'.[78] By making this connection between Abel and Cauchy, Mittag-Leffler directly linked him with the large transformations in analysis known as 'rigourisation' or 'arithmetisation', whose greatest protagonists were thus Gauss, Cauchy, Abel and Weierstrass. Abel was elevated to the highest level in nineteenth-century mathematical analysis, despite the fact that only a few of his publications dealt directly with 'rigourisation' and that his methods in the theory of elliptic functions still left ample room for major contributions from, for example, Bernhard Riemann (1826–1866) and Weierstrass.[79] Once again, the connection between Abel and Weierstrass was important for Mittag-Leffler.

Mittag-Leffler also cast Abel's mathematical practice and context in terms that resonate with the emerging 'modernity' in mathematics and in Scandinavia around 1900. For instance, he emphasized how travel and rapid communication through specialized journals had been important to Abel's mathematical life and how 'unlike many other mathematicians, Abel was well read in the works of others'.[80] Again, these matters were important to Mittag-Leffler himself, who had built his mathematical persona on his extensive travels on the Continent and his role as editor of one of the leading mathematical journals.

Continuing this line of argument, Mittag-Leffler emphasized how Abel recognized that the learned languages of the nineteenth century were to be French and German, not Latin,[81] and deplored the fact that, nonetheless, academic institutions in Norway and elsewhere would continue for

---

78 Mittag-Leffler, *Niels Henrik Abel* (1903), 134.
79 On Abel's contributions to the rigorization of analysis, see also Henrik Kragh Sørensen, 'Louis Olivier: A mathematician only known through his publications in Crelle's *Journal* during the 1820s', *Centaurus* 48/3 (2006), 201–31; and idem, 'Throwing Some Light on the Vast Darkness that is Analysis: Niels Henrik Abel's Critical Revision and the Concept of Absolute Convergence', *Centaurus* 52/1 (2010), 38–72.
80 Mittag-Leffler, *Niels Henrik Abel* (1903), 72.
81 Ibid.

decades to emphasize the dead language over the living. Indirectly he thus emphasized Abel's dual commitment to French and German culture and mathematics: another theme that had been important for Mittag-Leffler himself, during his own European tour in the aftermath of the Franco-Prussian war of 1872. It would, as world events unfolded in the twentieth century, become ever more important to him.[82]

Combining the modernity within contemporary mathematics with Abel's work on 'rigourisation', Mittag-Leffler suggested he had grasped what Weierstrass would later express by comparing mathematics to poetry:[83] the free, creative component of mathematics and its purity, setting aside its practical applications.[84] This, too, related to the increasing 'modernity' of the mathematical community in the decades around 1900.[85] The reorientation towards an autonomous discipline founded on the arithmetisation favoured by Weierstrass could be traced back to Abel:

---

82 Having his biography of Abel translated into French in 1907 is also indicative of Mittag-Leffler's international aspirations. At the same time it allowed him to enter a French discourse on Abel that had been reopened by the French edition of Holst's biography as well as the publication of an extensive biography by Pesloüan: Elling Holst, 'Niels Henrik Abel. Introduction historique a sa correspondance', in Holst et al., *Niels Henrik Abel*; Charles Lucas de Pesloüan, *N.-H. Abel: sa vie et son œuvre* (Paris: Gauthier-Villars, 1906). This important theme of the reception of the biographies abroad is, unfortunately, beyond the scope of the present article.

83 Mittag-Leffler, *Niels Henrik Abel* (1903), 134.

84 Often, and in particular after a famous radio lecture by Hilbert in 1930 (David Hilbert, 'Naturerkennen und Logik', in *David Hilbert. Gesammelte Abhandlungen* (3 vols, Berlin: Verlag von Julius Springer, 1930), vol. 3, pp. 378–87, at p. 387), such arguments are tied to Jacobi's famous dictum that 'the one and only purpose of mathematics is to honour the human spirit' (Jacobi to Legendre, Koenigsberg, 2 July 1830: Legendre, Adrien-Marie and Carl Gustav Jacob Jacobi, 'Correspondance mathématique entre Legendre et Jacobi', *Journal für die reine und angewandte Mathematik* 80 (1875), 205–79 (reproduced in *C.G.J. Jacobis Gesammelte Werke*, vol. 1, pp. 385–461), at 271–3, quote on 272–3; see also Herbert Pieper, ed., *Korrespondenz Adrien-Marie Legendre – Carl Gustav Jacob Jacobi. Correspondance mathématique entre Legendre et Jacobi* (Stuttgart, Leipzig: B.G. Teubner, 1998), 161).

85 See Jeremy Gray, *Plato's Ghost: The Modernist Transformation of Mathematics* (Princeton and Oxford: Princeton University Press, 2008) for an interesting and thorough discussion of modernity and modernism in mathematics.

Abel was the first great mathematician to openly and without concern throw down his mask. To him, mathematics holds its ideal within itself. Its object is numbers. Its purpose is thought about numbers.[86]

The claim that Abel indeed was an open adherent of the new style in mathematics as Mittag-Leffler understood it, and that he was the first 'great mathematician' to declare such a position, is debatable if not downright controversial.[87] Yet it allowed Mittag-Leffler to project his own values about mathematics into the past, and to connect them to the most famous and important mathematicians of the nineteenth century.

Mittag-Leffler and the material legacy after Abel

Mittag-Leffler was quick to assert himself as an intellectual heir to Abel's legacy: Holmboe was worthy of a prominent place in the history of mathematics for recognizing Abel, though he himself was not an important mathematician.[88] Bjerknes's biography was a 'valuable compilation of sources' and Holst's an 'excellent and empathic' biography,[89] though Mittag-Leffler downplayed his understanding of Abel's mathematics, instead pointing to the 'truly scientific presentation' in Sylow's contribution to the *Festskrift*.[90] Thus, by implication, Mittag-Leffler's biography was important because he was the

---

86   Mittag-Leffler, *Niels Henrik Abel* (1903), 140: the sentences that concluded the biography.
87   For an analysis pointing to the novelty of Abel's mathematics in a transition from a formula-centred towards a more concept-centred approach to mathematics, see for example Henrik Kragh Sørensen, 'Exceptions and counterexamples. Understanding Abel's comment on Cauchy's Theorem', *Historia Mathematica* 32/4 (2005), 453–80; idem, 'Representations as means and ends: Representability and habituation in mathematical analysis during the first part of the nineteenth century', in Bart Van Kerkhove, ed., *New Perspectives on Mathematical Practices: Essays in Philosophy and History of Mathematics* (New Jersey: World Scientific, 2009), 114–37; and idem, 'Throwing some light'.
88   Mittag-Leffler, *Niels Henrik Abel* (1903), 69–70.
89   Ibid., 132, 129.
90   Ibid., 129; Sylow's analysis was published as Ludvig Sylow, 'Abels Studier og hans Opdagelser', in Holst, Størmer and Sylow, *Festskrift*.

only one – except for Sylow – really to understand Abel's mathematics, making him almost uniquely equipped to assess Abel's life and production.

But Mittag-Leffler possessed more than just the intellectual assets for writing Abel's life; he also had privileged access to documents and information. For instance, he took a great interest in preserving the only extant portrait of Abel, which he had published in the first issue of *Acta Mathematica*. In 1903 it was in the private possession of one of Abel's relatives, in a deteriorating condition; reading between the lines of the biography, it is evident that Mittag-Leffler would have liked to take over the responsibility of preserving it.[91] Most spectacularly, he unearthed and purchased a lost and hitherto unpublished manuscript of a continuation of Abel's *Recherches*, which he published as part of a celebratory issue of *Acta Mathematica* in 1902.[92] This special volume also included contributions from fifty leading mathematicians describing recent developments originating in Abel's research. And through his many international contacts and personal acquaintances, Mittag-Leffler had amassed material which put him in a privileged position for writing the biography. For instance, he could draw on oral communications from Weierstrass, Boeck and Joseph Liouville (1809–1882).[93]

Together with his editorial and educational experience, which he drew on in passing judgements about events in Abel's life, all of this enabled Mittag-Leffler to claim to be Abel's true heir. This almost came to a confrontation in connection with the centennial celebrations in Norway in 1902, when Norwegian voices were afraid that Mittag-Leffler was in the process of hijacking or 'Swedifying' the event and the hero. Uproar was caused when he presented the special volume of *Acta Mathematica* in something like competition with the official *Festskrift*, and then invited all the foreign guests to his home outside Stockholm following the official event. It took considerable diplomatic efforts from Bjørnson and Mittag-

---

91 Mittag-Leffler, *Niels Henrik Abel* (1903), 74.
92 Niels Henrik Abel, 'Recherches sur les fonctions elliptiques (Second mémoire)', *Acta Mathematica* 26 (1902), 3–42; Gösta Mittag-Leffler, 'Un mémoire d'Abel', *Acta Mathematica* 26 (1902), 1–2.
93 Mittag-Leffler, *Niels Henrik Abel* (1903), 77, 75, 83.

Leffler to calm public opinion and to argue that he was the 'best friend of Norway in Stockholm'.[94]

Conclusions

I have shown how Mittag-Leffler used an excursion into the history of mathematics to position and fashion himself as the legitimate heir to Abel's mathematics. This he did using his insider-perspectives as a research mathematician and as an editor and centrally-placed figure in the international network, drawing on his sources and experiences no previous biographer of Abel had used.

But Mittag-Leffler also took an outsider's perspective by elevating himself above the patriotic connotations and importance of Abel's biography, at a time when national sentiments were being spurred in Norway. This allowed him to advance, instead, his own agenda of an international and mathematical community as the proper setting for understanding Abel's life. The progress of mathematics was, to Mittag-Leffler, more important than issues of national pride and identity.

In these ways – and others – Mittag-Leffler sought to use his biographical essay on Abel to position himself as a continuation of Abel's life and work both in personal and mathematical terms. He shared with Abel some of the difficulties of the system – or so he perceived – and he found in Abel a hero and a beacon for pure, autonomous, modern mathematics. If mathematics in Scandinavia had been backward before Abel, it was changing a century later, when a strong generation of Scandinavian mathematicians were making themselves felt on the international circuit, not least through Mittag-Leffler's contributions, entrepreneurship and networking.

---

94   For more on this incident, see Sørensen, 'Niels Henrik Abel's Political and Professional Legacy'.

JACQUELINE STEDALL
THE QUEEN'S COLLEGE, OXFORD

Thomas Harriot (1560–1621): history and historiography

The bare facts

The first entry for Thomas Harriot in the historical record is his registration at St Mary's Hall, Oxford, on 20 December 1577. His age was given as 17, which makes his year of birth 1560. He was also described as 'plebeian', and from Oxfordshire, but his background is otherwise unknown.

During his time in Oxford or soon afterwards Harriot became acquainted with Walter Raleigh, for whom he worked during the early 1580s. In 1585 Harriot sailed to America on a voyage funded by Raleigh, and lived for a year on Roanoke Island inside the outer banks of what is now North Carolina. The book he wrote on his return, *A briefe and true report of the new found land of Virginia* (1588), a public relations exercise to encourage further expeditions to the region, is all that he published in his lifetime.

Sometime in the early 1590s Harriot came under a second patron, Henry Percy, ninth earl of Northumberland. After the gunpowder plot in 1605, Percy was arrested and kept in the Tower of London for 16 years but Harriot continued to reside at the earl's London home, Syon House, on the Thames in Middlesex. There he worked on a range of mathematical and scientific subjects: optics, ballistics, alchemy, algebra, geometry, navigation, astronomy.

Harriot was in regular contact with a small but close-knit group of friends and colleagues: Walter Warner (a contemporary of Harriot's at Oxford, later employed for many years by Henry Percy as curator of his library and scientific instruments); Robert Hues (a contemporary of Harriot's at Oxford, later tutor to Percy's sons); Nathaniel Torporley (at Oxford shortly after Harriot); William Lower (Percy's son-in-law and steward of his Welsh estates); John Protheroe (successor to Lower after the latter's death in 1615); Thomas Aylesbury (possibly an Oxford friend of Protheroe's).

Harriot died in 1621, leaving behind him some 8,000 sheets of manuscript writings. Ten years after his death some of his work on algebra was published by Walter Warner in a book entitled *Artis analyticae praxis* (1631). These are the bare facts. The rest, we might say, is history.

Contemporary evidence and hearsay

Harriot's reputation was already well established within his own circle of friends during the 1590s. Gabriel Harvey in 1593 named 'Digges, Hariot, or Dee' as examples of 'profounde mathematicians';[1] a year later Robert Hues described him as 'most skilled in mathematics and general philosophy';[2] Nathaniel Torporley wrote of Harriot in 1602 as 'foremost in all kinds of learning.'[3]

In 1610, Harriot's friend William Lower urged him to publish his work on the non-circular motion of planets, on 'weight in water', and on algebra:

---

1 Gabriel Harvey, *Pierces supererogation or a new prayse of the old asse* (London: Iohn VVolfe, 1593), 190.
2 '[...] Thoma Harioto Matheseos & vniuersae Philosophiae peritissmo [...]', Robert Hues, *Tractatus de globis et eorum usu* (London: Thomas Dawson, 1594), 111.
3 '[...] in omni eruditonis varietate principem virum Thomam Hariotum, [...]', Nathaniel Torporley, *Diclides coelometricae* (London: Felix Kingston, 1602), sig. A2^v.

Thomas Harriot (1560–1621)

al these were your deues and manie others that I could mention [...] Onlie let this remember you, that it is possible by to much procrastination to be prevented in the honor of some of your rarest inventions and speculations.[4]

Harriot never did publish, however, and so the ground for other people's speculations became increasingly fertile, even in Harriot's lifetime and certainly after his death.

Four years after Harriot died, Henry Briggs, by then Savilian Professor of geometry at Oxford, wrote to Kepler that publication of Harriot's papers was expected shortly.[5] A year before the *Praxis* was published George Hakewill published another letter from Briggs listing 'the most observable inventions of moderne Mathematicians unknowne to the Ancients'; one of them, according to Briggs, was the understanding of spherical triangles which, he claimed, was first taught by Harriot.[6] Now Harriot had never published on spherical triangles so Briggs can only have heard this from one of Harriot's friends. His source was almost certainly Nathaniel Torporley, who held some of Harriot's papers during the 1620s and who is known to have been in touch with Briggs at that time. Thus the rumours and hearsay began.

It is clear from contemporary correspondence that Aylesbury and Warner planned to publish further papers after the *Praxis*.[7] In the 1630s as in the 1620s, stories about the papers circulated in intellectual circles. Samuel Hartlib noted in his *Ephemerides* in 1635 that 'Mr Warner hase all Harriot's manuscripts' and was 'setting some of them forth'; four years

---

4 Lower to Harriot, 6 February 1610, cited in Stephen Peter Rigaud, *Supplement to Dr Bradley's miscellaneous works with an account of Harriot's astronomical papers* (Oxford: Clarendon Press, 1833), 42.
5 'Cum propediem expectemus et exoptemus ipsius auctoris librum posthumum' [Since we may expect and hope for a posthumous book from that author any day]: Briggs to Kepler, 10 March 1625, in Johann Kepler, ed. Christian Frisch, *Joannis Kepleri astronomi opera omnia* (8 vols, Frankfurt and Erlangen: Heyder & Zimmer, 1858–72), vol. 4, 661–2.
6 'primus docuit peritissimus Geometra Thomas Hariottus' [first taught by the most skilled geometer Thomas Harriot]: George Hakewill, *An apologie of the power and providence of God in the government of the world* (second edition, Oxford: William Turner for Robert Allott, 1630), 263–4.
7 Aylesbury to Percy, 5 April 1632: London, British Library, Add. MS 4409, fol. 87.

later he recorded that his own friend and protégé John Pell was working on some of Harriot's problems and that 'Sir Thomas Aylesbury promised to let him have Harriot's papers'.

Twelve years later, in 1651, Aylesbury was living in exile in Antwerp. His London library, 'wherein were many rare and curious books', had meanwhile been plundered,[8] but Harriot's manuscripts had been saved because Aylesbury had kept them with him. When Sir Charles Cavendish visited Aylesbury in the autumn of 1651, they discussed some of Harriot's work and Cavendish, seeking Pell's assistance, informed Pell that Aylesbury could 'send you the originall'.[9]

That is the last sighting we have of the papers before Aylesbury died in exile in 1657. In the late 1650s he, or someone on his behalf, must have deposited them privately with the descendants of Henry Percy at Petworth House in Sussex.

Some collective memory of their existence lingered on. During the 1660s the Royal Society instigated two searches for Harriot's writings, one amongst Aylesbury's family papers in 1662, and a second amongst John Protheroe's in 1669, but to no avail. The consensus was that they were lost. In his *Treatise of algebra*, written during the early 1670s, John Wallis wrote of them: 'they are not publick: nor do I know in whose hands they are; if extant; nor whether they are ever like to see the light'.[10] In 1677, he wrote a preamble in the Savile Library copy of Harriot's *Praxis*, to much the same effect: 'There were many other very worthy pieces of Mr Harriot's doing, left behind him, & well worth the publishing [...] But in who's hands they now are, or whether they be since perished, I cannot tell.'[11]

8   London, British Library, MS Rawlinson B.158, fol. 153.
9   Cavendish to Pell, [26 September]/6 October 1651, London, British Library, Add. MS 4278, fols 321–2; reproduced in Noel Malcolm and Jacqueline Stedall, *John Pell (1611–1685) and his correspondence with Sir Charles Cavendish: the mental world of an early modern mathematician* (Oxford: Oxford University Press, 2005), 584.
10 John Wallis, *A treatise of algebra both historical and practical* (London: John Playford for Richard Davis, 1685), 198.
11 John Wallis, in Oxford, Bodleian Library, MS Savile O.9. For Wallis's inscription in full see John W. Shirley, *Thomas Harriot: a biography* (Oxford: Clarendon Press, 1983), 10–11.

The papers may have been to all intents and purposes lost, but Wallis knew something about them all the same. In his *Treatise of algebra* he wrote about Harriot's algebra extensively and it is clear that he had some inside knowledge of it. He imitated, for instance, the distinctive double zero, oo, that Harriot always used in his manuscripts to preserve dimensions, as in $aa - ba + ca - bc = oo$; this was not, however, to be found in the *Praxis*. Later, in 1693, Wallis admitted that his knowledge of Harriot's work came from Pell, who had urged him to write his account.[12] Pell was almost pathologically averse to having his name appear in print and had probably insisted that Wallis should not mention him, but he died in 1685, the year the *Treatise of algebra* was published, and after that Wallis probably felt free to name his sources. Pell was indeed an excellent authority: he had first engaged with Harriot's mathematics in the 1630s and had been personally acquainted with Aylesbury, Warner, and Cavendish. With his death first-hand knowledge of Harriot's work came to end. Any account that followed could be based on no more than the few published papers edited by Warner.

Controversies

Wallis devoted some 80 pages of his *Treatise of algebra* to explaining Harriot's innovations and 'improvements' in algebra. Unfortunately his account contained several statements that were either mathematically or historically controversial, the repercussions of which have continued to the present day.

One of the comments by Wallis that provoked protest was on the types of roots of equations that Harriot allowed:[13]

---

12 John Wallis, *Opera mathematica* (3 vols, Oxford: at the Sheldonian Theatre, 1693–99), vol. 2, following the preface.
13 Wallis, *A treatise*, 128.

> Beside the *Positive* or *Affirmative* Roots, (which he doth, through his whole Treatise, more especially pursue, as the principal and most considerable:) He takes in also the *Negative* or *Privative* Roots; which by some are neglected.

It is true that in his manuscripts Harriot took account of negative roots. When Warner edited the relevant pages for the *Praxis*, however, he reverted to the more usual conventions of the time, listing the positive roots but ignoring the negative. Wallis was thus referring to what he knew Harriot had written in the manuscripts, not to what appeared in the *Praxis*. It is not surprising, since Wallis did not state his sources or his evidence, that his statement was met with blank disbelief.

A similar claim about imaginary roots strained credulity even further:[14]

> And of such imaginary Roots, we find Mr Harriot particularly to take notice (in the Solution of Cubick Equations) [...]

Again, there are imaginary roots in plenty in the manuscripts, but none in the *Praxis*, where the reader finds imaginary roots not taken notice of at all, but on the contrary dismissed as 'inexplicable'.

Wallis must have known that his claims could not be supported by anything that readers could see in the *Praxis*, but when challenged directly resorted to a curious manner of dealing with the problem. He pointed to precisely the pages and equations where negative or imaginary roots *would* have appeared if Warner had not edited them out.[15] In fact he repeatedly encouraged readers to make their own inferences, rather than providing direct evidence of what Harriot had actually said. Thus his account abounds with phrases like "'Tis manifest' or 'Which appears by a bare inspection' or 'It follows naturally', seventeenth-century equivalents of the modern 'Clearly ...'. Almost all the twenty-five 'Improvements of Algebra' that Wallis attributed to Harriot are actually Wallis's interpretations of Harriot's work, examples of what he considered the astute reader could easily deduce.

---

14 Ibid., 134.
15 Wallis to Morland, 12 March 1688, in Wallis, *Opera mathematica*, vol. 1, pp. 209–10.

Unfortunately, it is all too easy to treat Harriot's writings in this way because he himself made so few explicit statements. If Wallis read more into Harriot's writings than is strictly to be found there, he did so partly because he was better informed than most of his contemporaries about the contents of the manuscripts. But he also did it for less commendable reasons, namely, to establish Harriot's priority and superiority over his French contemporary Descartes.

There are many disparaging references to Descartes in Wallis's account. This is one of the first and most pernicious:[16]

> [Harriot] hath made very many advantageous improvements in this Art; and hath laid the foundation on which Des Cartes (though without naming him,) hath built the greatest part (if not the whole) of his Algebra or Geometry. Without which, that whole Superstructure of Des Cartes (I doubt) had never been.

In a letter to Samuel Morland in 1688, Wallis denied that he had ever called Descartes a plagiarist,[17] but nevertheless insisted there and in all his further comments on the matter that Harriot's work had preceded that of Descartes by about forty years. This was true, but did not address the more relevant question of whether Descartes knew of the contents of the *Praxis* before he published *La géométrie* in 1637. Indeed, Jean Beaugrand had supposed as early as 1638 that Descartes did, but Descartes denied it, referring to Harriot as 'Henriotti', and saying that he would have liked to see his book because he had been told that it contained ideas similar to his own.[18] The *Praxis* was available in Paris in the 1630s and its contents could easily have been relayed to Descartes even if he had not seen it for himself. Even a brief description of Harriot's ideas about factorizing polynomials could

---

16   Wallis, *A treatise*, 126.
17   See note 15.
18   'c'est Henriotti [...] l'avois eu desir de voir ce livre, a cause qu'on m'avoit dit qu'il contenoit un calcul pour la geometrie, qui estoit fort semblable au mien.' [this 'Henriotti' [...] I wanted to see his book, because someone told me that it contains a calculus for geometry, which is very similar to mine.] Descartes [to Huygens], [December 1638], René Descartes, ed. Charles Adam and Paul Tannery, *Oeuvres de Descartes* (12 vols, Paris: Cerf, 1897–1910), vol. 1, pp. 479–80.

have been quite enough to set Descartes thinking along the same lines. On the other hand, it is not at all uncommon for mathematicians to work on similar ideas independently and almost simultaneously, an explanation we must accept if we are to take Descartes' denial at face value.

We simply do not know the truth of the matter now, and nor did anyone then, but that did not prevent people from expressing their opinions, invariably in line with their nationality. In 1689 the French mathematician Jean Prestet wrote with scorn:[19]

> Et lorsque Monsieur Wallis, une peu trop jaloux de la gloire que la France s'est acquis dans les Mathématiques, vient renouveller cette accusation ridicule, on est en droit de ne le point croire, puis qu'il parle sans preuve.
>
> [And when Monsieur Wallis, a little too jealous of the glory with which France has acquitted herself in mathematics, has just renewed this ridiculous accusation, one is right not to believe it at all, since he speaks without proof.]

Nevertheless, the English accusations lingered on. As late as 1839 the English historian Henry Hallam wrote:[20]

> It must be owned that, independently of the suspicions of an unacknowledged plagiarism of what others had thought before him, which unfortunately hang over all the writings of Descartes, he has taken to himself the whole theory of Harriot on the nature of equations in a manner which, if it is not a remarkable case of simultaneous invention, can only be reckoned a very unwarranted plagiarism.

Writing in the mid-eighteenth-century, Jean Montucla, in his four-volume *Histoire des mathématiques* was scathing about Wallis's account:[21]

---

19   Jean Prestet, *Nouveaux élémens des mathématiques ou principes généraux de toutes les sciences qui ont les grandeurs pour objet. Seconde éd. plus ample et mieux digérée* (Paris: A. Pralard, 1689), vol. 2, Preface.
20   Henry Hallam, *Introduction to the literature of Europe in the fifteenth, sixteenth and seventeenth centuries* (4 vols, London: J. Murray, 1837–1839), vol. 4, p. 18.
21   Jean E. Montucla, *Histoire des Mathématiques* (Paris: Jombert, 1758), vol. 2, p. 82.

> Comment excuserons-nous M. *Wallis* qui, nous donnant un Traité historique de l'Algebre, semble avoir à peine jetté les yeux sur texte autre Analiste qu'*Harriot* [...] Qui pourra même ne pas rire en voyant ce zélé restaurateur de la gloire d'Harriot.
>
> [How are we to excuse Monsieur Wallis who, giving us a historical Treatise of Algebra, seems scarcely to have cast his eyes on the text of any Analyst except Harriot. [...] Who would even not laugh on seeing this zealous restorer of the glory of Harriot.]

Montucla did, however, point fairly to Harriot's true achievement, his insight into the multiplicative structure of equations.[22]

> La découverte fondamentale d'Harriot, celle qui l'illustre parmi les analystes, consiste à avoit remarqué que toutes les equations d'ordres supérieurs sont des produits d'équations simples.
>
> [Harriot's fundamental discovery, which makes him famous among analysts, consists of having observed that all equations of higher order are products of simple equations.]

Montucla was not completely accurate here because Harriot had not made any claims for *all* equations of higher order, but he had examined a very large number, up to degree four. His insight that an equation with zero on the right could be re-written as a product of linear (or sometimes quadratic) factors was precisely what Descartes had also briefly stated, and which had thus given rise to Wallis's later accusations.

Charles Hutton too, in his article on algebra in his *Mathematical and philosophical dictionary* of 1796, pointed to what he thought was Harriot's finest achievement:[23]

> [Harriot] shewed the universal generation of all the compound or affected equations, by the continual multiplication of so many simple ones. [...] from which many of the most important properties [of equations] have since been deduced.

---

22 Ibid., 78–9.
23 Charles Hutton *A mathematical and philosophical dictionary* (2 vols, London: J. Johnson and G.G. & J. Robinson, 1795–96), vol. 1, p. 91 (s.v. 'Algebra').

In the centuries that followed, this claim, clearly stated by both Hutton and Montucla, somehow disappeared from the sight of historians. Thus, John Stillwell in his *Mathematics and its history* in 2000 attributed to Descartes 'the theorem that a polynomial $p(x)$ with value 0 when $x = a$ has a factor $(x - a)$'. Even if we ignore the anachronisms of Stillwell's notation, we cannot overlook his re-writing of the historical record.[24] Like Hutton and Montucla before him he has attributed too much to early seventeenth-century writers who dealt not so much in general theorems as in examples (in Harriot's case many, in Descartes' case just one); but worse, we may note that Harriot's name has here disappeared from the account altogether.

For most of the eighteenth century controversies about Harriot's attainments or priority were fought on sterile ground, because the only evidence by then available was Warner's muddled and incomplete edition of some of Harriot's algebra in the *Praxis*.[25] It might be thought that the eventual rediscovery of the manuscripts in 1784 would have put an end to all arguments, but it was not so simple. The papers do not reveal their import easily to casual readers, and rather than laying to rest old quarrels they merely fomented new ones.

The papers were rediscovered at Petworth in 1784 by Count von Bruhl, who had married the widow of the second earl of Egremont. He passed them to his son's Austrian tutor, Franz Xaver Zach, who offered to edit some of them, alongside a biography of Harriot, for the Clarendon Press at Oxford. Thus Hutton wrote in the 1790s that 'it is with pleasure I can announce that they are in a fair train to be published'.[26] Zach, for his part, was convinced that the manuscripts would vindicate Harriot against the criticisms of the French. But he also introduced a new element of controversy. 'It has not hitherto been known', he wrote, 'that Harriot was an

---

24 John Stillwell, *Mathematics and its history* (New York and London: Springer, 1989), 97.
25 As late as 1928 Florian Cajori still thought it worthwhile to re-evaluate the *Praxis* without reference to the manuscripts.
26 Hutton, *Dictionary*, vol. 1, p. 586.

eminent astronomer'.²⁷ Zach, an astronomer himself, was undoubtedly excited to discover Harriot's telescopic observations of sunspots and the satellites of Jupiter in 1610, but ought to have been more cautious before recklessly implying Harriot's priority over Galileo. Harriot's observations were in fact slightly later than Galileo's, though in the case of sunspots only by a few weeks.²⁸ The modern historian would see in this an intriguing case of parallel developments in Italy and England; indeed Harriot continues to this day to be described from time to time as 'the English Galileo', but for better argued reasons than Zach was able to offer.²⁹

As for publication, Zach, like earlier editors of Harriot, had taken on a more formidable task than he could manage. In 1794 he simply sent the unedited papers to the Delegates of the university press. They in turn sought advice from Abram Robertson, who at the time was standing in for Henry John Smith, the Oxford Savilian Professor of geometry. Robertson was unenthusiastic:³⁰

> These papers [...] are in no point of view fit for publication. The greatest part of them consist of detached and unfinished explanations of the authors which he read; begun, according to all appearance, with the design of satisfying his own mind upon the subject before him, and dropped abruptly as soon as this satisfaction was obtained.

Robertson was critical too of Harriot's mathematical style:

> No first principles are laid down; due arrangement is overlooked; and the demonstrations, often defective, are expressed in a kind of algebraic shorthand. [...]

In short, he concluded that Harriot had never intended the papers for publication and that 'it would be injurious to his reputation to print them'.

---

27 From a piece published by Zach in Berlin in 1788 and later translated by him into English; cited in Rigaud, *Supplement*, 58.
28 See Rigaud, *Supplement*, 57–61, for Zach's claims, and 19–41, for Rigaud's lengthy discussion of them.
29 See in particular Matthias Schemmel, *The English Galileo: Thomas Harriot's work on motion as an example of preclassical mechanics* (2 vols, Dordrecht: Springer, 2008).
30 Robertson's report is given in full in Rigaud, *Supplement*, 61–3.

From an editorial point of view one has every sympathy with Robertson's assessment. To historians, however, Harriot's notes on other authors, his algebraic shorthand, and his demonstrations, even if defective, are all of considerable interest.

Reinforced by Zach's exaggerations and Hutton's promises, rumours continued to circulate as late as the 1820s that the Press at Oxford was committed to publishing the papers but was dragging its feet. In defence of both the Press, to which he was a Delegate, and of Robertson, his predecessor as Savilian professor of geometry, Stephen Peter Rigaud began a fresh examination of the manuscripts, and arrived at a more measured assessment than Zach's:[31]

> To help in establishing the fair fame of Harriot would be a source of the highest gratification to me. He was not only a countryman, but a distinguished member of that university to which I have myself the honour of belonging. If truth hath obliged me to shew that his astronomical observations are not of that high character which they have been supposed to possess, he can well spare what does not belong to him. There is no degradation in being second to Galileo.

One might think that Rigaud's careful judgement would have brought the matter to an end, but emotions for or against Harriot are not so easily put to rest. In 1963, Johannes Lohne, one of the best twentieth-century Harriot scholars, wrote a paper entitled 'The fair fame of Thomas Harriott: Rigaud versus Baron von Zach', in which he claimed that:[32]

> The English professors from Harriott's university did their utmost to conjure the inopportune Harriott back again into relative obscurity. Unfortunately they succeeded.

This was certainly not true of Rigaud; rather than conjuring Harriot back into obscurity he filled many sheets with notes on Harriot's mathematics

---

31 Ibid., 51.
32 Johannes A. Lohne, 'The fair fame of Thomas Harriott: Rigaud versus Baron von Zach', *Centaurus* 8 (1963), 69–84.

and astronomical observations, and many more with items of biographical information on Harriot's various acquaintances.[33]

After Rigaud's death in 1839 interest in Harriot waned once again; there was after all very little more to say. The most serious long-term consequence of Zach's intervention was the disturbance of the papers. Most of those Zach selected for publication were returned to Oxford and are now retained at Petworth. Some may have suffered a worse fate. The first half of an important letter from Lower to Harriot, for instance, quoted (but misattributed) by Zach is now lost.[34] Then in August 2010 a fragment headed 'Magneticall Experimente', containing a few mathematical notes in Harriot's handwriting, turned up in a US auction room and was sold to a dealer for several thousand dollars. We do not know its journey to that point, nor what else might be out there. The papers untouched by Zach were meanwhile gifted (in 1810) by the third earl of Egremont to the British Museum, where they were bound without reordering into eight large volumes. Thus the bulk of the papers remain to this day divided between the British Library and Petworth.

The construction of mythologies

Controversies around Harriot up to the end of the nineteenth century tended to focus on the subjects first raised by Wallis: his recognition of negative roots or his priority over Descartes (or, in Zach's perception, over Galileo). In 1900, however, a small book changed the course of Harriot studies and paved the way for some of the speculations that characterized twentieth-century assessments. Its author was Henry Stevens, an American employed for most of his life in the British Museum, who in 1877 had founded the Hercules Club with the aim of publishing material

---

33   Oxford, Bodleian Library, MSS Rigaud 35, 9, 61, 56.
34   The letter is reproduced in full in Rigaud, *Supplement*, 42–3 and 44–5.

relating to early Anglo-American history. Stevens' first project was to reproduce Harriot's *Briefe and true report* with a biographical introduction. Unfortunately, Stevens died before he could complete the work, but the biographical material was published by his son in 1900 under the breathless title *Thomas Hariot the mathematician the philosopher and the scholar developed chiefly from dormant materials with notices of his associates including biographical and bibliographical disquisitions upon the materials of the history of 'ould Virginia'*. It is usually known as *Thomas Hariot and his associates*. The original is rare because only 162 copies were printed; the only copy in Oxford is in the library of Rhodes House among other material relating to early colonial adventures.

Stevens collected together as much information as he could find on Harriot and his friends. His key discovery was Harriot's will at the Archdeaconry Court in London, where no-one had previously thought of searching for it. The will is a crucial document for historians of Harriot's mathematics because it names the people Harriot entrusted with his books and papers and enables us to begin to piece together the story of their circulation in the seventeenth century.

Stevens' style of writing, however, set the adulatory tone that was to follow in so much twentieth-century literature about Harriot. Here is just one example, Stevens' description of the friendship between Torporley and Harriot:[35]

> Torporley, who had entered St Mary's Hall the year Hariot graduated, and who during his travels abroad had served two years as private secretary or amanuensis to Francis Vieta, the great French Mathematician, but who had since become a disciple of the greater English Mathematician, thus admiringly speaks of his new master.

The paragraph is factually incorrect because Torporley was in fact at Christ Church. Otherwise it contains only unsupported opinions. Whether Harriot was a 'greater' mathematician than Viète must be a matter of judgment, but I think few historians would be prepared to argue for such a

---

35 Henry Stevens, *Thomas Hariot the mathematician the philosopher and the scholar* [...] (London: Privately printed at the Chiswick press, 1900), 101.

claim. As to Torporley's role of 'disciple' to Harriot as 'master', the facts speak otherwise: it was Torporley who brought Viète's work to Harriot in what, as we may surmise from the one surviving letter between them, both regarded as a friendship between equals.

Unfortunately, speculation based on half-truths continued. In 1903 Arthur Acheson devised the name 'The school of night' for a group of poets and writers around Raleigh, supposedly opposed to Shakespeare. The concept was taken up again by Frances Yates in 1636 and, in the same year, in a study entitled *The school of night* by Muriel Bradbrook, who wrote: 'Ralegh was the patron of the school; Harriot, a mathematician of European reputation, its master.'[36] A 'European reputation' would be hard to prove for a man whose mathematics was known only to his personal acquaintances and who for twenty-five years rarely travelled beyond a fifteen-mile stretch of the Thames. As for being master of any 'school', there is no evidence for it whatsoever.

It is true that both Raleigh and Harriot were from time to time accused of atheism. In Harriot's case such accusations seem to have been provoked mainly by his view, first suggested to him by the Indians of north America, that the earth might be very much older than biblical chronology indicated. Even his friend Torporley was dismayed by Harriot's claim that *nihil ex nihil fit* (nothing is created from nothing). But such a view was not incompatible with the belief in divine law also expressed by Harriot: 'Credo in deum omnipotentem. Credo medicinam ab illo ordinatur. Credo medico tanquam illius ministro.' [I believe in God the all-powerful. I believe in the art of medicine ordained by Him. I believe in the physician as His minister.][37] These are hardly the words of an atheist.

In recent years, several fanciful portraits of Harriot have found their way into popular accounts. It would be hard to do worse than Muriel

---

36 Muriel Bradbrook, *The school of night: a study in the literary relationships of Sir Walter Raleigh* (Cambridge: Cambridge University Press, 1936), 8.
37 London, British Library, Add. MS 6789, fol. 446^v, reproduced in Shirley, *Thomas Harriot*, 436.

Rukeyser's *The traces of Thomas Harriot* (1971). The book itself is not worth quoting but I cannot resist giving D.T. Whiteside's view of it:[38]

> The sober English printing lacks the visual extravagance of the American *editio princeps*, whose dust-jacket proclaims as its by-line that 'He burst from history into legend' and carries on its rear a giant close-up of Miss Rukeyser's left eye.

Then more recently we had Giles Milton's *Big chief Elizabeth* (2000), in which he described Harriot with his 'algebraic intellect' as:[39]

> a mathematical conjurer, a wizard, who was also blessed with an imagination that spawned such side-shoots that he had already [in his early twenties] solved some of the most impenetrable scientific conundrums of the day.

The problem with most of these accounts is that they come from authors with little understanding of Harriot's major life's work, his mathematics. Cecily Tanner, who first began publishing on Harriot in 1963, was better placed than most to rectify this omission. There are few traces of solid mathematical research in her published papers, however, which tend to dwell on minor matters like Harriot's inequality signs or his rules for multiplication. Her more general assessments were often flawed, as in her extraordinary claim that the *Praxis* contains only 'advanced arithmetic with a minimum of algebraic foundations, nothing like algebra in its own right'.[40]

Nevertheless, Tanner's enthusiasm for Harriot did much to revive interest in the mathematical and scientific contents of his papers, to the extent that during the 1970s a Harriot Seminar was established. A group of its members pledged to edit the entire corpus of Harriot's papers,[41] but as so often before the muddle and disorder of the papers defeated the effort.

---

38 Derek Thomas Whiteside, 'In search of Thomas Harriot' (essay review), *History of Science* 13 (1975), 61–70, note 5.
39 Giles Milton, *Big chief Elizabeth: how England's adventurers gambled and won the New World* (London: Hodder & Stoughton, 2000), 47–8.
40 Rosalind Cecilia H. Tanner, 'Thomas Harriot as mathematician: a legacy of hearsay', *Physis* 9 (1967), 235–47.
41 See the comment on this by Derek Thomas Whiteside in 'In search of Thomas Harriot', 67. Key members of the seminar in its early days were Gordon Batho, Alistair

Instead, photocopies of the British Library manuscripts were deposited in the university libraries of Durham and Oxford.

The biography of Harriot published by John Shirley in 1983 finally cast a more sober and realistic light on Harriot and his life.[42] In particular, Shirley thoroughly demolished the myth of the 'three magi' (Harriot, Warner, Hues) kept by the 'wizard earl' in the Tower of London; instead Shirley saw a small group of friends supporting each other as well as they could through the difficult and sometimes dangerous circumstances of their everyday lives. Shirley's balanced and thoroughly researched biography did Harriot studies an enormous service. Its shortcoming remains, however, that of so many other twentieth-century writings on Harriot, a lack of engagement with his mathematics. It is to that final stage of Harriot historiography that we turn next.

Examining the mathematics

The earliest detailed examination of Harriot's mathematics in the twentieth century was Jon Pepper's reconstruction of Harriot's calculations of meridional parts, the corrections needed at each latitude if one is to steer a correct compass course rather than spiralling in to one of the earth's poles along a rhumb line.[43] These calculations demonstrate Harriot's understanding of stereographic projection, his ability to calculate lengths of spiral arcs, and his sophisticated use of interpolation.

Gradually, other aspects of Harriot's mathematics came under the spotlight. The most comprehensive published survey of the mathematical

---

Crombie, Ivor Grattan-Guinness, John North, Jon Pepper, David Quinn, John Roche, John Shirley, Cecily Tanner and David Quinn.

42 Shirley, *Thomas Harriot*.
43 Jon V. Pepper, 'Harriot's calculations of the meridional parts as logarithmic tangents', *Archive for History of Exact Sciences* 4 (1968), 359–413.

papers to date was made by the Norwegian mathematics teacher Johannes Lohne who visited England regularly during his summer vacations in the 1960s and 1970s to work on Harriot's manuscripts,[44] and who correlated the topics he found in them with those listed in an inventory compiled shortly after Harriot's death.[45] Lohne took a particular interest in Harriot's work on collisions and projectiles.[46] He wrote very little, however, on Harriot's algebra, which he passed over with a few brief examples chosen to refute the old accusations that Harriot did not work with negative or imaginary roots.[47] Nor did he investigate in any detail Harriot's relationship to Viète.[48] Only in the last ten years has there begun to be a more thorough exploration of Viète's crucial influence on Harriot's mathematical thought.[49]

Meanwhile, by the 1990s, Muriel Seltman had begun to work on the neglected topic of Harriot's algebra. Her starting point was the old question of what kind of roots Harriot allowed, to which end she examined the various equations and solutions to be found in the manuscripts. She made many insightful observations on Harriot's notation and page layout, and the relationship of the manuscript material to the contents of the *Praxis*. Her chief legacy to Harriot studies is the English translation of the *Praxis* that she and Robert Goulding published in 2007.

44 Reinhard Siegmund-Schultze, 'Johannes Lohne (1908–1993) revisited: documents for his life and work, half a century after his pioneering research on Harriot and Newton', *Archives Internationales d'histoire des sciences* 60 (2010), 569–96.
45 Johannes A. Lohne, 'Essays on Thomas Harriot. III. A survey of Harriot's scientific writings', *Archive for History of Exact Sciences* 20 (1979), 265–312.
46 Johannes A Lohne., 'Essays on Thomas Harriot. I. Billiard balls and laws of collision', *Archive for History of Exact Sciences* 20 (1979), 189–229; idem, 'Essays on Thomas Harriot. II. Ballistic parabolas', *Archive for History of Exact Sciences* 20 (1979), 230–64.
47 Lohne, 'Essays on Thomas Harriot. III', 295–8.
48 Ibid., 288–9, 296–7, 305–6.
49 Jacqueline Stedall, 'Notes made by Thomas Harriot on the treatises of François Viète', *Archive for History of Exact Sciences* 62 (2008), 179–200; Janet Beery and Jacqueline Stedall, *Thomas Harriot's doctrine of triangular numbers: the 'magisteria magna'* (Zürich: European Mathematical Society, 2009).

My own work on Harriot's manuscripts was inspired by a document by Torporley, in which he listed the pages he thought should be included in an edition of Harriot's algebra.[50] In editing those 140 pages, I chose at the time to transcribe and translate Harriot's script; indeed I was advised by a respected scholar to do so, otherwise, he said, 'one might just as well present the reader with the originals'. I now think it would have been much better to present the reader with the originals. Translation, it can be argued, is necessary in order to put Harriot's sparse notes and comments into English for the sake of those who do not read Latin. Transcription, on the other hand, inevitably distorts the original. Not least, text printed line-by-line from left to right, cannot capture Harriot's use of page-space to connect webs of interrelated ideas. Besides, modern typesetting is clean and clear and fails to produce an imitation or even an impression of a handwritten document.

When Janet Beery and I came to edit Harriot's *Magisteria magna*, we decided, for the reasons outlined above, to offer facsimiles with facing commentary.[51] Matthias Schemmel, working on Harriot's studies of projectile motion, independently adopted a similar approach, and at the same time added meta-textual commentary in the form of diagrams to show how groups of pages or individual sheets are interrelated.[52]

The similarity of our vision for the manuscripts has led Matthias Schemmel and me to embark upon a new publication project, this time electronically.[53] By making use of the space and flexibility of web publication we aim not only to make the entire corpus of the manuscripts freely

50   London, Lambeth Palace Library, Sion College MS Arc L.40.2/L.40, fols 35–54ᵛ. This document was preserved at Sion College after Torporley's death there in 1632; it was transferred with other Sion College holdings to Lambeth Palace in 1996. For discussion of it and a reproduction of the first page see Jacqueline Stedall, *The greate invention of algebra: Thomas Harriot's treatise on equations* (Oxford: Oxford University Press, 2003), 24–6.
51   Beery and Stedall, *Thomas Harriot's doctrine of triangular numbers*.
52   Schemmel, *The English Galileo*.
53   *Collection of Source Texts on Thomas Harriot*, European Cultural Heritage Online (ECHO): <http://echo.mpiwg-berlin.mpg.de/content/scientific_revolution/harriot>.

available but also to offer translations, commentary, and cross-referencing. Our aspiration is to make Harriot's writings not only physically available but also intellectually accessible.

Thus, for the fourth time in four centuries, there is a rumour abroad that Harriot's papers are to be published. This time, however, the rumour is becoming reality: the contents of the British Library volumes are already imported into European Cultural Heritage Online (ECHO), established and hosted by the Max-Planck Institut für Wissenschaftsgeschichte in Berlin. The next stage will be to classify the manuscripts under key headings: arithmetic, algebra, geometry, mechanics, navigation, optics, and so on (with some papers inevitably appearing in more than one section). Within each section, smaller groups of related pages will be identified and analysed, together with rough work pertaining to them. In this way, papers that have been scattered over different parts of the collection will be electronically re-connected, often from more than one volume, sometimes from more than one library. The initial commentaries will come from a small group of scholars who are already familiar with aspects of Harriot's work, but we hope that as the work progresses it will accommodate new contributors and new audiences. Harriot could not have foreseen such publication; nevertheless I like to think that the wishes expressed in his will are at last being carried out.

The prospects are exciting but new methods also present new problems. One of the most serious is securing electronic publications into the future. Will digital reproductions survive as robustly as the printed *Praxis*, or even the manuscripts, through the technological changes that inevitably lie ahead? We must hope and suppose that they will. Then come the intellectual questions. Will full publication of the material finally lead scholars to a more unified view of Harriot's achievements or, as so often in the past, only to more, and more diverse, interpretations of his work? And in the light of the historiography outlined in this chapter I am forced to ask myself another question: will historians of the future look back and see our present attempts to do justice to Harriot as yet another well-intentioned but misguided effort, determined by but also limited by the transient conditions of time and place?

Bibliography

Abel, Niels Henrik, 'Recherches sur les fonctions elliptiques', *Journal für die reine und angewandte Mathematik* 2/2 (1827), 101–81 (reproduced in Abel, *Oeuvres Complètes* (1881), vol. 1, pp. 263–388).
—— 'Recherches sur les fonctions elliptiques (Second mémoire)', *Acta Mathematica* 26 (1902), 3–42.
—— 'Recherches sur les fonctions elliptiques', *Journal für die reine und angewandte Mathematik* 3/2 (1828), 160–90 (reproduced in Abel, *Oeuvres Complètes* (1881), vol. 1, pp. 263–388).
—— et al., 'Breve fra og til Abel', in Holst, Størmer and Sylow, *Festskrift*.
—— ed. Bernt Michael Holmboe, *Oeuvres Complètes de N.H. Abel, mathématicien, avec des notes et développements* (2 vols, Christiania: Chr. Gröndahl, 1839).
—— ed. Ludvig Sylow and Sophus Lie, *Oeuvres Complètes de Niels Henrik Abel* (2 vols, Christiania: Grøndahl, 1881).
Acheson, Arthur, *Shakespeare and the rival poet* (London: J. Lane, 1903).
Alexander, Amir R., 'Tragic Mathematics: Romantic Narratives and the Refounding of Mathematics in the Early Nineteenth Century', *Isis* 97/4 (2006), 714–26.
Arbuthnot, John, *An Essay on the Usefulness of Mathematical Learning* (Oxford: at the Theater, for Anth. Peisley, 1701).
Aubin, David and Charlotte Bigg, 'Neither Genius nor Context Incarnate: Norman Lockyer, Jules Janssen and the Astrophysical Self', in Söderqvist, *Scientific Biography*, 51–70.
Bailey, Nathan, *An universal etymological English dictionary* (London: J Darby and others, 1728).
Bailly, Jean Sylvain, *Histoire de l'astronomie ancienne: depuis son origine jusqu'à l'établissement de l'école d'Alexandrie* (Paris: Frères Debure, 1775).
—— *Histoire de l'astronomie moderne: depuis la fondation de l'école d'Alexandrie, jusqu'à l'époque de M.D.CC.XXX* (Paris: Frères Debure, 1779–82).
—— *Traité de l'astronomie indienne et orientale: ouvrage qui peut servir de suite à l'histoire de l'astronomie ancienne* (Paris: Debure l'aîné, 1787).
Baily, Francis, *An account of the Rev. John Flamsteed* (London: William Clowes, 1835).
Ball, Walter William Rouse, *A history of the study of mathematics at Cambridge* (Cambridge: Cambridge University Press, 1889).

—— *A short account of the history of mathematics* (London: Macmillan, 1888; 4th edition, 1908).

Barrow, Isaac, *Lectiones XVIII, Cantabrigiae in Scholis publicis habitae; in quibus opticorum phaenomenωn genuinae rationes investigantur, ac exponuntur* (London: W. Godbid, 1669).

Beaujouan, G., 'Lagrange et Montucla', *Revue d'Histoire des Sciences* 3 (1950), 128–32.

Beeley, Philip, '"Un de mes amis". On Leibniz's relation to the English mathematician and theologian John Wallis', in Pauline Phemister and Stuart Brown, eds, *Leibniz and the English-Speaking World* (Dordrecht: Springer, 2007), 63–81.

—— 'Experiment, Induktion, Hypothese. Leibniz' Auseinandersetzung mit dem wissenschaftlichen Nachlaß des Joachim Jungius um 1678', *Studia Leibnitiana* (forthcoming).

—— and Christoph J. Scriba, 'Wallis, Leibniz und der Fall von Harriot und Descartes. Zur Geschichte eines vermeintlichen Plagiats im 17. Jahrhundert', *Acta Historica Leopoldina* 45 (2005), 115–29.

—— and Christoph J. Scriba, 'Disputed Glory. John Wallis and some questions of precedence in seventeenth-century mathematics', in Hartmut Hecht *et al.*, eds, *Kosmos und Zahl. Beiträge zur Mathematik- und Astronomiegeschichte, zu Alexander von Humboldt und Leibniz* (Stuttgart: Steiner Verlag, 2008), 275–99.

Beery, Janet, and Jacqueline Stedall, *Thomas Harriot's doctrine of triangular numbers: the 'magisteria magna'* (Zürich: European Mathematical Society, 2009).

Bell, Eric Temple, *Men of mathematics* (New York: Simon and Schuster, 1937).

Bentley, Richard, *Eight Sermons, preached at the Hon. Robert Boyle's Lecture, in the year mdcxcii* (Oxford: Clarendon Press, 1809).

Bertoloni Meli, Domenico, *Equivalence and priority: Newton versus Leibniz, including Leibniz's unpublished manuscripts on the Principia* (Oxford: Clarendon Press, 1993).

Bertrand, Joseph, review of Bjerknes, *Niels-Henrik Abel*, in *Bulletin des sciences mathématiques* (second series) 9 (1885), 190–202.

[Biot, Jean-Baptiste], 'Newton (Isaac)', in L.G. Michaud, ed., *Biographie universelle, ancienne et moderne*, vol. 31 (Paris: Michaud Frères, 1822), 127–94.

Biot, Jean-Baptiste, *Life of Newton* [translated by Howard Elphinstone] (London: s.n., 1829).

—— 'Revue de *The Life of Isaac Newton*', *Journal des savants* (1836), 156–66, 205–23, 641–58.

—— and Felix Lefort, eds, *Commercium epistolicum* (Paris: Mallet-Bachelier, 1856).

Birch, Thomas, *The History of the Royal Society of London for Improving of Natural Knowledge* (4 vols, London: for A. Millar, 1756–7).

Bjerknes, Carl Anton, *Niels Henrik Abel. En skildring af hans liv og videnskabelige virksomhed*, published as an appendix to *Nordisk tidsskrift för vetenskap, konst och industri* (Stockholm: P.A. Norstedt & Söner, 1880).

—— *Niels-Henrik Abel. Tableau de sa vie et des on action scientifique* (Paris: Gauthier-Villars, 1885).

Bjørnson, Bjørnstjerne, *Samlede værker. Mindeutgave* (5 vols, Kristiania and Kjøbenhavn: Gyldendalske Boghandel & Nordisk Forlag, 1910–1911).

Boistel, Guy, Jérôme Lamy and Colette Le Lay, eds, *Jérôme Lalande (1732–1807): une trajectoire scientifique* (Rennes: Presses universitaires de Rennes, 2010).

Bossut, Charles, *Essai sur l'histoire générale des mathématiques jusqu'en 1784.* (Paris: Louis, 1802).

Bougainville, Louis Antoine de, *Traité du calcul intégral* (Paris: Desaint & Saillant, 1754).

Bradbrook, Muriel, *The school of night: a study in the literary relationships of Sir Walter Raleigh* (Cambridge: Cambridge University Press, 1936).

Brewster, David, *The life of Sir Isaac Newton* (London: John Murray, 1831).

—— *Memoirs of the life, writings, and discoveries of Sir Isaac Newton* (2 vols, Edinburgh: Thomas Constable & Co., 1855).

Brunel, G., review of Bjerknes, *Niels-Henrik Abel*, in *Bulletin des sciences mathématiques* (second series), 9 (1885), 141–53.

Butler, William, *Arithmetical questions, on a new plan* (second edition, London: for the author, 1795).

Byrne, James Steven, 'A Humanist History of Mathematics? Regiomontanus's Padua Oration in Context', *Journal for the History of Ideas* 67 (2006), 41–61.

Caine, Barbara, *Biography and History* (Basingstoke: Palgrave Macmillan, 2010).

Cajori, Florian, *A history of mathematics* (New York: Macmillan, 1894).

—— 'A revaluation of Harriot's Artis Analyticae Praxis', *Isis* 11 (1928), 316–24.

Cantor, Geoffrey N., 'Between rationalism and romanticism: Whewell's historiography of the inductive sciences', in Fisch and Schaffer, *William Whewell*, 67–86.

Carver, Jonathan, trans. J.E. Montucla, *Voyage dans les Parties Interieures de l'Amerique Septentrionale* (Paris: Chez Pissot, 1784).

Christie, John R.R., 'Sir David Brewster as an historian of science', in A.D. Morrison-Low and John R.R. Christie, eds, *Martyr of science: Sir David Brewster 1781–1868* (Edinburgh: Royal Scottish Museum, 1984), 53–6.

Clark, David H. and Stephen P.H. Clark, *Newton's Tyranny: The Suppressed Scientific Discoveries of Stephen Gray and John Flamsteed* (W.H. Freeman and Co., 2001).

Clark, Samuel, 'Arithmetic', in Temple Henry Croker, *The complete dictionary of arts and sciences* (3 vols, London: for the authors, 1764), vol. 1.

Cocker, Edward, *A Treatise of Arithmetic* (new edition, Edinburgh: for E. Wilson, 1765).

Collection of Source Texts on Thomas Harriot, European Cultural Heritage Online (ECHO): <http://echo.mpiwg-berlin.mpg.de/content/scientific_revolution/harriot>.

Collins, John, 'A Letter from Mr. John Collins to the Reverend and Learned Dr. John Wallis Savilian Professor of Geometry in the University of Oxford, giving his thoughts about some Defects in Algebra', *Philosophical Transactions* 159 (20 May 1684), 575–82.

Condorcet, Jean-Antoine-Nicolas de Caritat de, *Tableau historique du progrès de l'esprit humain* (Paris: Agasse, 1795).

Crepel, Pierre and Alain Coste, 'Jean-Étienne Montucla, *Histoire des mathématiques*, second edition (1799–1802)', in Ivor Grattan-Guinness, ed., *Landmark Writings in Western Mathematics, Case Studies 1640–1940* (Amsterdam: Elsevier, 2005), 292–302.

Cusa, Nicholas of, ed. J. Hofmann and J.E. Hofmann, *Die mathematischen Schriften* (Hamburg: Meiner Verlag, 1952).

Dati, Carlo, *Lettera a Filalethi di Timauro Antiate della vera Storia della Cicloide e della famosissima esperienza dell' Argento vivo* (Florence: Insegna della Stella, 1663).

Dauben, Joseph W. and Christoph J. Scriba, eds, *Writing the history of mathematics: Its historical development* (Basel: Birkhäuser, 2002).

De Morgan, Augustus, 'Newton', in Charles Knight, ed., *The Cabinet portrait gallery of British worthies* (12 vols, London: Charles Knight & Co., 1845–49), vol. 11, pp. 78–117 (reprinted in De Morgan, *Essays on the life and work of Newton*, 3–63).

—— 'A short account of some recent discoveries in England and Germany relative to the controversy on the invention of fluxions', *Companion to the almanac for 1852*, 5–20 (reprinted in De Morgan, *Essays on the life and work of Newton*, 67–101).

—— 'On a point connected with the dispute between Keill and Leibnitz about the invention of fluxions', *Philosophical transactions of the Royal Society* 136 (1846), 107–9.

—— 'On the additions made to the second edition of the Commercium Epistolicum', *Philosophical magazine* (third series) 32 (1848), 446–56.

—— review of Brewster's *Memoirs of the life of Sir Isaac Newton*, *North British Review* 23 (1855), 307–38 (reprinted in De Morgan, *Essays on the life and work of Newton*, 119–82).

—— ed. Sophia Elizabeth De Morgan and Arthur Cowper Ranyard, *Newton: his friend and his niece* (London: Elliot Stock, 1885) (excerpts reprinted in Iliffe, Keynes and Higgitt, *Early biographies of Isaac Newton*, vol. 2, pp. 289–337).

―― ed. Philip E.B. Jourdain, *Essays on the life and work of Newton* (Chicago and London: The Open Court Publishing Company, 1914).

De Morgan, Sophia Elizabeth, *Memoir of Augustus De Morgan* (London: Longmans, Green & Co., 1882).

Delambre, Jean Baptiste Joseph, *Rapport historique sur les progrès des sciences mathématiques depuis 1789: et sur leur état actuel* (Paris: de l'Imprimerie impériale, 1810).

Descartes, René, ed. Charles Adam and Paul Tannery, *Oeuvres de Descartes* (12 vols, Paris: Cerf, 1897-1910).

Dobbs, Betty Jo T., *The foundations of Newton's alchemy or 'The Hunting of the Greene Lyon'* (Cambridge: Cambridge University Press, 1975).

Donne, Benjamin, *Mathematical essays; being essays on vulgar and decimal arithmetic* (second edition, London: W. Johnston, P. Davey and B. Law, 1758).

Dumont, Simone, preface by Jean Claude Pecker, *Un astronome des lumières: Gerome Lalande* (Paris: Vuibert, 2007).

Edleston, J., *Correspondence of Sir Isaac Newton and Professor Cotes* (London: John W. Parker, 1850).

Fara, Patricia, *Newton: the Making of Genius* (London: Macmillan, 2002).

Fauvel, John, Raymond Flood, Michael Shortland and Robin Wilson, eds, *Let Newton Be!* (Oxford: Oxford University Press, 1988).

Fenn, Joseph, *A New and Complete System of Algebra* (Dublin: Alex. McCulloh, 1775?).

The first principles of arithmetic. By a late teacher of mathematics in the Royal Navy. (London: C. Forster, 1792).

Fisch, Menachem and Simon Schaffer, eds, *William Whewell: A composite portrait* (Oxford: Clarendon Press, 1991).

Fontenelle, Bernard le Bovier de, *The Life of Sir Isaac Newton. With an Account of his Writings* (London: James Woodman and David Lyon, 1728) (English translation of 1727 'Eloge du Chevalier Newton').

Gascoigne, John, 'From Bentley to the Victorians: The rise and fall of British Newtonian natural theology', *Science in context* 2 (1988), 219-56.

Gjertsen, Derek, *The Newton Handbook* (London and New York: Routledge & Kegan Paul, 1986).

Gleick, James, *Isaac Newton* (London: Fourth Estate, 2003).

Goulding, Robert, *Defending Hypatia: Ramus, Savile, and the Renaissance rediscovery of mathematical history* (Dordrecht: Springer, 2010).

Grattan-Guinness, Ivor, *Convolutions in French Mathematics, 1800-1840* (Basel: Birkhäuser, 1990).

Gray, Jeremy, *Plato's Ghost: The Modernist Transformation of Mathematics* (Princeton and Oxford: Princeton University Press, 2008).

Gröning, Johann, *Historia cycloidis, contra Pascalium, mathematicum Gallum* (Hamburg: Liebezeit, 1701).

Guicciardini, Niccolò, *The development of Newtonian calculus in Britain 1700–1800* (Cambridge: Cambridge University Press, 1989).

—— '"Gigantic implements of war": images of Newton as a mathematician', in Robson and Stedall, *Handbook*, 707–35.

Gunther, Robert William Theodore, *Early Science in Oxford*, vol. 12: *Dr. Plot and the Correspondence of the Philosophical Society of Oxford* (Oxford: Oxford University Press, 1939).

Hakewill, George, *An apologie of the power and providence of God in the government of the world* (second edition, Oxford: William Turner for Robert Allott, 1630).

Hall, A. Rupert, *Isaac Newton: Eighteenth Century Perspectives* (Oxford: Oxford University Press, 1999).

—— *Philosophers at war: The quarrel between Newton and Leibniz* (Cambridge: Cambridge University Press, 1980).

Hallam, Henry, *Introduction to the literature of Europe in the fifteenth, sixteenth and seventeenth centuries* (4 vols, London: J. Murray, 1837–1839).

Halley, Edmund, 'An Historical Account of the Trade Winds, and Monsoons, Observable in the Seas between and near the Tropicks, with an Attempt to Assign the Phisical Cause of the Said Winds', *Philosophical Transactions* 183 (July, August and September 1686), 153–68.

Hanks, Lesley, *Buffon avant l'Histoire naturelle* (Paris: Presses Universitaires de France, 1966).

Hansen, Signe Lindskov, 'The Programmatic Function of Biography: Readings of Nineteenth- and Twentieth-Century Biographies of Niels Stensen (Steno)', in Söderqvist, *Scientific Biography*, 135–53.

Hansteen, Christopher, 'Niels Henrik Abel', *Illustreret Nyhedsblad* 11/9–10 (2–9 March 1862), 37–8, 41–2.

The Hartlib Papers on CD-ROM, Sheffield, Humanities Research Institute (second edition, 2002).

Harvey, Gabriel, *Pierces supererogation or a new prayse of the old asse* (London: Iohn VVolfe, 1593).

Heard, John, 'The Evolution of the Pure Mathematician in England, 1850–1920', unpublished PhD dissertation. University of London, 2004.

Hestmark, Geir, *Vitenskap og nasjon. Waldemar Christopher Brøgger 1851–1905* (Oslo: H. Aschehoug & Co., 1999).

Bibliography

Hevelius, Johannes, *Mercurius in sole visus Gedani, anno Christiano MDCLXI* (Danzig: Reiniger 1662).
Higgitt, Rebekah, 'Why I don't FRS my tail: Augustus De Morgan and the Royal Society', *Notes and Records of the Royal Society* 60 (2006), 253–9.
—— *Recreating Newton: Newtonian biography and the making of nineteenth-century history of science* (London: Pickering & Chatto, 2007).
Hilbert, David, 'Naturerkennen und Logik', in *David Hilbert. Gesammelte Abhandlungen* (3 vols, Berlin: Verlag von Julius Springer, 1930), vol. 3, pp. 378–87.
Hofmann, Joseph E., *Leibniz in Paris, 1672–76* (Cambridge: Cambridge University Press, 1974).
Holmboe, Bernt Michael, 'Necrolog. Kort Fremstilling af Niels Henrik Abels Liv og videnskabelige Virksomhed', *Magazin for Naturvidenskaberne* (1829), 334–54.
Holst, Elling, 'Mathematik', in Henrik Jæger and Otto Anderssen, eds, *Illustreret norsk literaturhistorie. Videnskabernes litteratur i det nittende aarhundrede* (4 vols, Kristiania: Hjalmar Biglers Forlag, 1896), vol. 4, pp. 68–101.
—— 'Niels Henrik Abel. Historisk indledning til hans efterladte breve', in Holst, Størmer and Sylow, *Festskrift*.
—— 'Niels Henrik Abel. Introduction historique à sa correspondance', in Holst *et al*., *Niels Henrik Abel*.
—— *et al*., eds, *Niels Henrik Abel. Memorial publié à l'occasion du centenaire de sa naissance* (Kristiania: Jacob Dybwad etc., 1902).
—— Carl Størmer, and Ludvig Sylow, eds, *Festskrift ved Hundredeaarsjubilæet for Niels Henrik Abels Fødsel* (Kristiania: Jacob Dybwad, 1902).
Hoquet, Thierry, 'History without Time: Buffon's Natural History as a Nonmathematical Physique', *Isis* 101 (2010), 30–61.
Houzel, Christian, 'The Work of Niels Henrik Abel', in Olav Arnfinn Laudal and Ragni Piene, eds, *The Legacy of Niels Henrik Abel. The Abel Bicentennial, Oslo, 2002* (Berlin: Springer, 2004), 21–177.
Hues, Robert, *Tractatus de globis et eorum usu* (London: Thomas Dawson, 1594).
Hutton, Charles, *A mathematical and philosophical dictionary* (2 vols, London: J. Johnson and G.G. and J. Robinson, 1795–6).
—— *A philosophical and mathematical dictionary* (2 vols, London: F.C. and J. Rivington, 1815).
Iliffe, Rob, 'A "connected system"? The snare of a beautiful hand and the unity of Newton's archive', in Michael Hunter, ed., *Archives of the Scientific Revolution: the Formation and Exchange of Ideas in Seventeenth-Century Europe* (Woodbridge: Boydell Press, 1998), 137–57.
—— Milo Keynes and Rebekah Higgitt, eds, *Early biographies of Isaac Newton, 1660–1885*: vol. 1, *Eighteenth-century Biography of Isaac Newton: the Unpublished*

Manuscripts and Early Texts; vol. 2, *Nineteenth-Century Biography of Isaac Newton: Public Debate and Private Controversy* (2 vols, London: Pickering & Chatto, 2006).

Jacobi, Carl Gustav Jacob, ed. Carl Wilhelm Borchardt, Karl Theodor Wilhelm Weierstrass and Alfred Clebsch, *C.G.J. Jacobis Gesammelte Werke* (8 vols, Berlin: G. Reimer, 1881–1891; reprinted New York: Chelsea Publishing Company, 1969).

Johnson, Samuel, *A dictionary of the English language* (second edition, 2 vols, London: J. and P. Knapton and others, 1755–56).

Keill, John, 'Epistola ad clarissimum virum Edmundum Halleium Geometriae Professorem Savilianum, de legibus virium centripetarum', *Philosophical Transactions of the Royal Society* 26 (1708), 174–88.

Kepler, Johann, ed. Christian Frisch, *Joannis Kepleri astronomi opera omnia* (8 vols, Frankfurt and Erlangen: Heyder & Zimmer, 1858–72).

Keynes, John Maynard, 'Newton, the man', in J.M. Keynes, ed. G. Keynes, *Essays in Biography* [...] *New Edition with Three Additional Essays* (London: Rupert Hart-Davis, 1951), 310–23.

Koenigsberger, Leo, *Zur Geschichte der Theorie der elliptischen Transcendenten in den Jahren 1826–29* (Leipzig: B.G. Teubner, 1879).

Lacroix, Sylvestre-François, *Traité du calcul différentiel et du calcul intégral* (Paris: Duprat, 1797–1800).

Laita, Luis M., 'Influences on Boole's logic: the controversy between William Hamilton and Augustus De Morgan', *Annals of science* 36 (1979), 45–65.

Lalande, Jérôme, review of vols 1 and 2 of Montucla, *Histoire* (1799–1802), *Magasin encyclopédique* 3 (1799), 256–9.

——*Bibliographie astronomique: avec l'histoire de l'astronomie depuis 1781 jusqu'à 1802* (Paris: de l'Imprimerie de la République, 1803).

Lampe, Emil, review of Mittag-Leffler, *Niels Henrik Abel*, in *Jahrbuch über die Fortschritte der Mathematik* 40.0020.01 (1908).

Landgren, Per, *Det aristoteliska historiebegreppet. Historieteori i renässansens Europa och Sverige* (Gothenburg: Göteborgs Universitet, 2008).

Le Blond, Auguste Savinien, *Notice historique sur la vie et les ouvrages de Montucla* (Versailles: Societé Libre d'Agriculture de Seine-et-Oise, 1800).

Le Clerc, Daniel, *Histoire de la médicine* (Amsterdam: G. Gallet, 1702).

Le Seur, Thomas and François Jacquier, *Elemens du Calcul Intégral* (Parma: Chez les Heritiers Monti, 1768).

Lee, Hermione, *Biography. A Very Short Introduction* (Oxford: Oxford University Press, 2009).

Legendre, Adrien-Marie and Carl Gustav Jacob Jacobi, 'Correspondance mathématique entre Legendre et Jacobi', *Journal für die reine und angewandte Mathematik* 80 (1875), 205–79 (reproduced in *C.G.J. Jacobis Gesammelte Werke*, vol. 1, pp. 385–461).

—— and —— ed. Herbert Pieper, *Korrespondenz Adrien-Marie Legendre – Carl Gustav Jacob Jacobi. Correspondance mathématique entre Legendre et Jacobi* (Stuttgart, Leipzig: B.G. Teubner, 1998).

Leibniz, Gottfried Wilhelm, 'De vera proportione circuli ad quadratum circumscriptum in numeris rationalibus expressa', *Acta eruditorum* (February 1682), 41–6.

—— 'Nova methodus pro maximus et minimus [...]', *Acta eruditorum* (October 1684), 467–73.

—— 'De geometria recondita et analysi indivisibilium et infinitorum [...]', *Acta eruditorum* (1686), 226–33.

—— review of Wallis, *Treatise of Algebra*, *Acta eruditorum* (June 1686), 283–9.

—— anonymous review of Newton's 'Enumeratio linearum tertii ordinis' and 'Tractatus de quadratura curvarum', *Acta eruditorum* (January 1705), 30–6.

—— ed. C.I. Gerhardt, *Leibnizens mathematische Schriften* (7 vols, Berlin and Halle: A. Asher and H.W. Schmidt, 1849–63; reprinted Hildesheim: Georg Olms, 1962).

—— ed. Prussian Academy of Sciences (and successors), *Sämtliche Schriften und Briefe* (Darmstadt: Reichl Verlag (and successors), 1923–).

Lewis, Clive Staples, *The Discarded Image: an introduction to medieval and Renaissance literature* (new edition, Cambridge: Cambridge University Press, 1994).

Lohne, Johannes A., 'The fair fame of Thomas Harriott: Rigaud versus Baron von Zach', *Centaurus* 8 (1963), 69–84.

—— 'Essays on Thomas Harriot. I. Billiard balls and laws of collision', *Archive for the History of Exact Science* 20 (1979), 189–229.

—— 'Essays on Thomas Harriot. II. Ballistic parabolas', *Archive for the History of Exact Science* 20 (1979), 230–64.

—— 'Essays on Thomas Harriot. III. A survey of Harriot's scientific writings', *Archive for the History of Exact Science* 20 (1979), 265–312.

'Mrs Lovechild' [Eleanor Fenn], *The Art of Teaching in Sport; Designed as a Prelude to a set of Toys, for enabling Ladies to Instill the Rudiments of Spelling, Reading, Grammar, and Arithmetic, under the Idea of Amusement* (London: John Marshall and Co., 1785).

Maccioni Ruju, P. Alessandra and Marco Mostert, *The life and times of Guglielmo Libri: scientist, patriot, scholar, journalist, and thief* (Hilversum: Verloren, 1995).

Maclaurin, Colin, *Treatise of Fluxions* (Edinburgh: Ruddimans, 1742).

Malcolm, Alexander, *A Treatise of Musick; Speculative, Practical, and Historical* (Edinburgh: for the author, 1721).

——*A new system of arithmetick, theorical and practical* (London: J. Osborn and others, 1730).

Malcolm, Noel, and Jacqueline Stedall, *John Pell (1611–1685) and his correspondence with Sir Charles Cavendish: the mental world of an early modern mathematician* (Oxford: Oxford University Press, 2005).

Mansfield, Elizabeth, 'Emilia Dilke: Self-Fashioning and the Nineteenth Century', in Marysa Demoor, ed., *Marketing the Author. Authorial Personae, Narrative Selves and Self-Fashioning, 1880–1930* (London: Palgrave Macmillan, 2004), 19–39.

Manuel, Frank, *A Portrait of Isaac Newton* (Cambridge, MA: Harvard/Belknap Press, 1968).

Milton, Giles, *Big chief Elizabeth: how England's adventurers gambled and won the New World* (London: Hodder & Stoughton, 2000).

Mittag-Leffler, Gösta, *En metod at komma i analytisk besittning af de Elliptiska Funktionerna* (Helsingfors: J.C. Frenckell & Son, 1876).

——'Un mémoire d'Abel', *Acta Mathematica* 26 (1902), 1–2.

——'Niels Henrik Abel', *Ord och bild: Illustrerad månadsskrift* 12 (1903), 65–85, 129–40.

——*Niels Henrik Abel* (Paris: La revue du mois, 1907).

——'Die ersten 40 Jahren des Lebens von Weierstraß', *Acta Mathematica* 39 (1923), 1–57.

——'Weierstraß et Sonja Kowalewsky', *Acta Mathematica* 39 (1923), 133–98.

Montucla, Jean E., *Histoire des Recherches sur la Quadrature du Cercle* (Paris: Jombert, 1754).

——*Histoire des Mathématiques* (Paris: Jombert, 1758).

——and Pierre Joseph Morisot-Deslandes, *Recueil de Pieces concernant l'Inoculation de la Petite Vérole, & propres à en prouver la sécurité & l'utilité* (Paris: Desaint & Saillant, 1756).

——revised by Joseph Jérôme L. de Lalande, *Histoire des Mathématiques* (4 vols, Paris: Henri Agasse, 1799–1802); reprinted with a preface by Charles Naux, Paris: Blanchard, 1960).

Morrison-Low, A.D. and John R.R. Christie, eds, *Martyr of science: Sir David Brewster 1781–1868* (Edinburgh: Royal Scottish Museum, 1984).

Nabonnand, Philippe, 'The Poincaré–Mittag-Leffler Relationship', *Mathematical Intelligencer* 21/2 (1999), 58–64.

Netto, Eugen, review of Bjerknes, *Niels-Henrik Abel*, in *Jahrbuch über die Fortschritte der Mathematik* 17.0014.04 (1885).

Newton, Isaac, *Opticks: or a treatise of the reflexions, refractions, inflexions and colours of light* (London: Sam. Smith and Benj. Walford, 1704).
—— *Commercium epistolicum D. Johannis Collins, et aliorum de analysi promota* (London: Pearson, 1712).
—— 'An account of the book entitled *Commercium epistolicum* [...]', *Philosophical Transactions of the Royal Society* 29 (1715), 173–224.
—— *Commercium epistolicum D. Johannis Collins, et aliorum de analysi promota* (second edition, London: J. Tonson and J. Watts, 1722).
—— *The Method of Fluxions and Infinite Series; with its Application to the Geometry of Curve-Lines* (London: by H. Woodfall, sold by J. Nourse, 1736).
—— trans. Georges-Louis Leclerc de Buffon, *La Methode des Fluxions et des Suites Infinies* (Paris: De Bure, 1740).
—— ed. S. Horsley, *Opera quae exstant omnia* (London: J. Nichols, 1779–1785).
—— ed. Herbert W. Turnbull, Joseph F. Scott, A. Rupert Hall and Laura Tilling, *The Correspondence of Isaac Newton* (7 vols, Cambridge: Cambridge University Press, 1967–77).
Nichols, John, *Illustrations of the literary history of the eighteenth century, consisting of authentic memoirs and original letters of eminent persons*, vol. 4 (London: John Nichols, 1822).
Nye, Mary Jo, Joan L. Richards and Roger H. Stuewer, eds, *The invention of physical science: Intesections of mathematics, theology and natural philosophy since the seventeenth century* (Dordrecht, Boston and London: Kluwer Academic Publishers, 1992).
Ore, Øystein, *Niels Henrik Abel. Et geni og hans samtid* (Oslo: Gyldendal Norsk Forlag, 1954).
—— *Niels Henrik Abel. Mathematician Extraordinary* (Minneapolis: University of Minnesota Press, 1957).
Osler, Margaret J., 'A hero for their times: early biographies of Newton', *Notes and Records of the Royal Society* 60 (2006), 291–305.
Ozanam, Jacques, ed. J.E. Montucla, *Récréations mathématiques et physiques* (Paris: Jombert, 1778).
Parke, Catherine Neal, *Biography. Writing Lives* (New York and London: Routledge, 2002).
Pepper, Jon V., 'Harriot's calculations of the meridional parts as logarithmic tangents', *Archive for History of Exact Sciences* 4 (1968), 359–413.
Pesloüan, Charles Lucas de, *N.-H. Abel: sa vie et son œuvre* (Paris: Gauthier-Villars, 1906).
Picutti, Ettore, 'La storia della matematica da Montucla a Cossali', *Regnum dei. collectanea theatina* 39/109 (1983), 179–85.

Playfair, John, 'Traité de Mechanique Celeste', *The Edinburgh Review* 22 (1808), 249–84.

Popper, Nicholas, '"Abraham, Planter of Mathematics": histories of mathematics and astrology in early modern Europe', *Journal for the History of Ideas* 67 (2006), 87–106.

Prestet, Jean, *Nouveaux élémens des mathématiques ou principes généraux de toutes les sciences qui ont les grandeurs pour objet. Seconde éd. plus ample et mieux digérée* (Paris: A. Pralard, 1689).

Ramus, Petrus, *Scholarum mathematicarum libri unus et triginta* (Basel: Episcopius, 1569).

Raphson, Joseph, *The history of fluxions* (London: William Pearson, 1715).

Rice, Adrian, 'Augustus De Morgan: historian of science', *History of science* 34 (1996), 201–40.

—— 'Inspiration or Desperation? Augustus De Morgan's appointment to the Chair of Mathematics at London University in 1828', *British Journal for the History of Science* 30 (1997), 257–74.

Richards, Joan L., 'Augustus De Morgan, the history of mathematics, and the foundations of algebra', *Isis* 78 (1987), 7–30.

—— 'Historical Mathematics in the French Eighteenth Century', *Isis* 97 (2006), 700–13.

Rigaud, Stephen Peter, *Supplement to Dr Bradley's miscellaneous works with an account of Harriot's astronomical papers* (Oxford: Clarendon Press, 1833).

—— *Historical Essay on the first publication of Sir Isaac Newton's Principia* (Oxford: Oxford University Press, 1838).

—— and Stephen Jordan Rigaud, eds, *Correspondence of scientific men of the seventeenth century* (2 vols, Oxford: Oxford University Press, 1841).

Robson, Eleanor and Jacqueline Stedall, eds, *The Oxford handbook of the history of mathematics* (Oxford: Oxford University Press, 2009).

Roche, John, 'Newton's *Principia*', in Fauvel et. al., *Let Newton Be!*, 43–61.

Roger, Jacques, trans. Sarah Lucille Bonnefoi, ed. L. Pearce Williams, *Buffon: a life in natural history* (Ithaca, NY and London: Cornell University Press, 1997).

Rothman, Tony, 'Genius and Biographers: the Fictionalization of Evariste Galois', *American Mathematical Monthly* 89 (1982), 84–106.

Sarton, George, 'Montucla (1725–1799): His Life and Works', *Osiris* 1 (1936), 519–67.

Savérien, Alexandre, *Dictionnaire universel de mathématique et de physique* (Paris: Jombert, 1753).

—— *Histoire critique du calcul des infiniment-petits: contenant la métaphysique et la théorie de ce calcul* (Paris: s.n., 1753).

—— *Histoire des progrès de l'esprit humain dans les sciences et dans les arts qui en dépendent* (Paris: chez Lacombe, 1766).

Savile, Henry, *Praelectiones tresdecim in principium Elementorum Euclidis Oxonii habitae MDCXX* (Oxford: Iohannes Lichfield, & Iacobus Short, 1621).

Schemmel, Matthias, *The English Galileo: Thomas Harriot's work on motion as an example of preclassical mechanics* (2 vols, Dordrecht: Springer, 2008).

Scriba, Christoph J., *Studien zu Mathematik des John Wallis (1616–1703)* (Wiesbaden: Steiner Verlag, 1966).

Seltman, Muriel, and Robert Goulding, *Thomas Harriot's* Artis analyticae praxis: *an English translation with commentary* (New York: Springer, 2007).

Shirley, John W., *Thomas Harriot: a biography* (Oxford: Clarendon Press, 1983).

Shortland, Michael and Richard Yeo, eds, *Telling Lives in Science: Essays on Scientific Biography* (Cambridge: Cambridge University Press, 1996).

Siegmund-Schultze, Reinhard, 'Johannes Lohne (1908–1993) revisited: documents for his life and work, half a century after his pioneering research on Harriot and Newton', *Archives Internationales d'histoire des sciences* 60 (2010), 569–96.

Sigurdsson, Skuli, 'Equivalence, pragmatic platonism, and discovery of the calculus', in Mary Jo Nye, Joan L. Richards and Roger H. Stuewer, eds, *The invention of physical science: Intesections of mathematics, theology and natural philosophy since the seventeenth century* (Dordrecht, Boston and London: Kluwer Academic Publishers, 1992), 97–116.

Slagstad, Rune, *De nasjonale strateger* (Oslo: Pax Forlag, 1998).

Smith, David E., 'John Wallis as a Cryptographer', *Bulletin of the American Mathematical Society* 24 (1917), 83–96 and 166–9.

—— 'Among my autographs. 6 [...] Montucla's history of mathematics', *American Mathematical Monthly* 28 (1921), 207–8.

—— 'Among my autographs. 25 [...] Montucla's closing years', *American Mathematical Monthly* 29 (1922), 253–55.

—— *History of mathematics* (2 vols, Boston: Ginn and Co., 1923–5).

—— *A Sourcebook in Mathematics* (New York and London: McGraw-Hill, 1929).

Söderqvist, Thomas, ed., *The History and Poetics of Scientific Biography* (Aldershot and Burlington, VT: Ashgate, 2007).

Sørensen, Henrik Kragh, 'Exceptions and counterexamples. Understanding Abel's comment on Cauchy's Theorem', *Historia Mathematica* 32/4 (2005), 453–80.

—— 'Louis Olivier: A mathematician only known through his publications in Crelle's *Journal* during the 1820s', *Centaurus* 48/3 (2006), 201–31.

—— 'Niels Henrik Abel's Political and Professional Legacy in Norway', in Reinhard Siegmund-Schultze and Henrik Kragh Sørensen, eds, *Perspectives on Scandinavian Science in the Early Twentieth Century* (Oslo: Novus forlag, 2006), 197–219.

—— 'Representations as means and ends: Representability and habituation in mathematical analysis during the first part of the nineteenth century', in Bart Van Kerkhove, ed., *New Perspectives on Mathematical Practices: Essays in Philosophy and History of Mathematics* (New Jersey: World Scientific, 2009), 114–37.

—— 'Throwing Some Light on the Vast Darkness that is Analysis: Niels Henrik Abel's Critical Revision and the Concept of Absolute Convergence', *Centaurus* 52/1 (2010), 38–72.

—— *The Mathematics of Niels Henrik Abel: Continuation and New Approaches in Mathematics During the 1820s*, Research Publications on Science Studies 11 (Department of Science Studies, University of Aarhus, 2010), <http://www.ivs.au.dk/reposs>.

Sprat, Thomas, *The History of the Royal-Society of London* (London: Thomas Roycroft for J. Martyn and J. Allestry, 1667).

Stedall, Jacqueline, 'Of our own Nation: John Wallis's account of mathematical learning in medieval England', *Historia mathematica* 28 (2001), 73–122.

—— *A Discourse concerning Algebra: English algebra to 1685* (Oxford: Oxford University Press, 2002).

—— *The greate invention of algebra: Thomas Harriot's treatise on equations* (Oxford: Oxford University Press, 2003).

—— 'Notes made by Thomas Harriot on the treatises of François Viète', *Archive for the History of Exact Science* 62 (2008), 179–200.

Stevens, Henry, *Thomas Hariot the mathematician the philosopher and the scholar* [...] (London: Privately printed at the Chiswick press, 1900) (often referred to as *Thomas Hariot and his associates*).

Stillwell, John, *Mathematics and its history* (New York and London: Springer, 1989).

Stubhaug, Arild, *Et foranskutt lyn – Niels Henrik Abel og hans tid* (Oslo: Aschehoug, 1996).

—— trans. Richard H. Daly, *Niels Henrik Abel and his Times. Called too soon by Flames Afar* (Berlin: Springer, 2000).

—— trans. Richard H. Daly, *The Mathematician Sophus Lie. It was the Audacity of My Thinking* (Berlin: Springer, 2002).

—— *Med viten og vilje: Gösta Mittag-Leffler (1846–1927)* (Oslo: Aschehoug, 2007).

—— trans. Tiina Nunnally, *Gösta Mittag-Leffler. A man of conviction* (Berlin and London: Springer, 2010).

A Supplement to the Athenian Oracle [...] (London: Andrew Bell, 1710).

Swerdlow, Noah M., 'Montucla's legacy: the history of the exact sciences', *Journal of the History of Ideas* 54 (1993), 299–328.

Bibliography

Sylow, Ludvig, 'Abels Studier og hans Opdagelser', in Holst, Størmer and Sylow, *Festskrift*.

Tanner, Rosalind Cecilia H., 'Thomas Harriot as mathematician: a legacy of hearsay', *Physis* 9 (1967), 235–47.

Theerman, Paul, 'Unaccustomed role: the scientist as historical biographer – two nineteenth-century portrayals of Newton', *Biography* (1985): 145–62.

Thomasen, Laura Søvsø and Henrik Kragh Sørensen, 'The Irony of Romantic Mathematics' (unpublished draft, September 2010).

Thomson, Thomas, *History of the Royal Society, from its institution to the end of the eighteenth century* (London: Robert Baldwin, 1812).

Toplis, John, 'On the Decline of Mathematical Studies, and the Sciences dependent upon them', *Philosophical Magazine* 20 (1805), 25–31.

Torporley, Nathaniel, *Diclides coelometricae* (London: Felix Kingston, 1602).

Turner, Laura E., 'A Man on a Mission: Mittag-Leffler's Efforts to Promote the Research Imperative to His Students at Stockholm's Högskola' (unpublished draft, May 2010).

Turnor, Edmund, *Collections for the History of the Town and Soke of Grantham* (London: W. Miller, 1806).

Van Brummelen, Glen, and Michael Kinyon, eds, *Mathematics and the Historian's Craft: The Kenneth O. May Lectures* (New York: Springer, 2005).

An universal history, from the earliest account of time to the present (7 vols, Dublin: George Faulkner, 1744–1745).

Vossius, Gerardus, *De quatuor artibus popularibus, de philologia et scientiis mathematicis, cui operi subjungitur, chronologia mathematicorum, libri tres* (Amsterdam: Bleau, 1650).

Wallis, John, *Mathesis universalis: sive, arithmeticum opus integrum* (Oxford: Leonard Lichfield for Thomas Robinson, 1657).

—— *Tractatus duo. Prior, de cycloide et corporibus inde genitis. Posterior, epistolaris; in qua agitur, de cissoide* (Oxford: Lichfields, 1659).

—— *Jeremiae Horroccii [...] Opera posthuma* (London: J. Martyn, 1673).

—— *A Proposal about Printing a Treatise of Algebra, Historical and Practical* (London: John Playford, 1683).

—— *A treatise of algebra both historical and practical* (London: John Playford for Richard Davis, 1685).

—— '[Account of] A Treatise of Algebra, both historical and practical', *Philosophical Transactions* 173 (22 July 1685), 1095–1105.

—— *Opera mathematica* (3 vols, Oxford: at the Sheldonian Theatre, 1693–99).

—— 'An extract of a letter from Dr Wallis, of May, 1697, concerning the Cycloeid known to Cardinal Cusanus about the year 1450; and to Carolus Bovillus about the year 1500', *Philosophical Transactions* 229 (June 1697), 561–6.

—— ed. Philip Beeley and Christoph J. Scriba, *The Correspondence of John Wallis (1616–1703)* (Oxford: Oxford University Press, 2003–).

Warwick, Andrew, *Masters of Theory: Cambridge and the Rise of Mathematical Physics* (Chicago and London: University of Chicago Press, 2003).

Westfall, Richard S., 'The changing world of the Newtonian industry', *Journal of the History of Ideas* 37 (1976), 175–84.

—— *Never at rest: A biography of Isaac Newton* (Cambridge: Cambridge University Press, 1980).

—— 'Newton and his biographer', in Samuel H. Baron and Carl Pletsch, eds, *Introspection in Biography: the Biographer's Quest for Self-Awareness* (Hillsdale, NJ and London: The Analytic Press, 1985), 175–89.

Whewell, William, *Newton and Flamsteed: Remarks on an article in number CIX of the Quarterly Review* (London: J. & J.J. Deighton, 1836).

—— *History of the inductive sciences* (3 vols, London: J.W. Parker, 1837).

Whiteside, Derek Thomas, ed., *The mathematical papers of Isaac Newton* (8 vols, Cambridge: Cambridge University Press, 1967–1981).

—— 'In search of Thomas Harriot' (essay review), *History of Science* 13 (1975), 61–70.

Yates, Francis, *A study of* Love's labour's lost (Cambridge: Cambridge University Press, 1936).

Yeo, Richard, 'Genius, method, and morality: Images of Newton in Britain, 1760–1860', *Science in Context* 2 (1988), 257–84.

—— *Defining science: William Whewell, natural knowledge, and public debate in early Victorian Britain* (Cambridge: Cambridge University Press, 1993).

Yoder, Joella G., *Unrolling Time: Christiaan Huygens and the mathematization of nature* (Cambridge: Cambridge University Press, 1988).

Zetterberg, J. Peter, 'The Mistaking of "the Mathematicks" for Magic in Tudor and Stuart England', *The Sixteenth Century Journal* 11/1 (1980), 83–97.

Notes on Contributors

PHILIP BEELEY is a fellow of Linacre College, Oxford, and joint editor (with Christoph J. Scriba) of the Correspondence of John Wallis. He has previously taught history of science and philosophy at the Technische Universität Berlin and the University of Hamburg, and was editor of the Academy Edition of Leibniz at the University of Münster. He is currently producing an edition of Wallis's *Treatise of Logic* and a volume of essays entitled *Mathesis metaphysica quadam. Interrelations between mathematics and metaphysics in Leibniz.*

NICCOLÒ GUICCIARDINI teaches history of science at the University of Bergamo. He is the author of *The Development of Newtonian Calculus in Britain, 1700–1800* (CUP, 1989), *Reading the Principia: the debate on Newton's mathematical methods for natural philosophy from 1687 to 1736* (CUP 1999), and *Isaac Newton on Mathematical Certainty and Method* (MIT 2009). He is co-Editor of *Historia Mathematica*.

REBEKAH HIGGITT has been Curator of History of Science and Technology at the Royal Observatory, Greenwich, and National Maritime Museum since 2008. She did her PhD at the Centre for the History of Science, Technology and Medicine at Imperial College London and was a postdoctoral researcher at the Institute of Geography, University of Edinburgh. She is the author of *Recreating Newton: Newtonian Biography and the Making of Nineteenth-Century History of Science* (Pickering Chatto, 2007).

ADRIAN RICE is Professor of Mathematics at Randolph-Macon College in Ashland, Virginia, where his research focuses on nineteenth- and early twentieth-century British mathematics. He is a two-time recipient of the Mathematical Association of America's Trevor Evans Award for outstanding expository writing.

HENRIK KRAGH SØRENSEN is Associate Professor in history of mathematics at the Centre for Science Studies at the University of Aarhus, Denmark. His research focuses on transformations in mathematics during the nineteenth century in the work of Niels Henrik Abel, and on the internationalisation of Scandinavian mathematics in the decades around 1900.

JACQUELINE STEDALL is a Senior Research Fellow in history of mathematics at The Queen's College, Oxford, and at the Mathematical Institute, Oxford. Her main research interests are in early modern English mathematics and the development of algebra. Her most recent book is *The history of mathematics: a very short introduction* for Oxford University Press.

BENJAMIN WARDHAUGH is a Post-doctoral Research Fellow at All Souls College, Oxford, where he works on mathematics and its uses in early modern Britain; he has a particular interest in the use of mathematics in early modern music theory. He is the author of *How to Read Historical Mathematics* (Princeton, 2010).

Index

Abel, Niels Henrik 6, 115
   centenary 6, 122–3, 137, 143
   clarity and elegance of 129, 134
   death of 121, 131
   European tour 120, 130, 133, 135–6, 141
   influence of 137–8
   and Jacobi 120, 122, 125, 127–30, 138
   letters of 122–3, 136
   as a 'modern' mathematician 139–42, 144
   and the Norwegian landscape 135–6
   and the Norwegian state 130–32
   portrait of 143
   poverty of 130–31
   and Weierstrass 140
   works of 121, 137
   see also Bjerknes, Carl Anton; Bjørnson, Bjørnstjerne; Holmboe, Bernt Michael; Holst, Elling; Mittag-Leffler, Gösta
Académie des sciences 52, 74, 82, 87–8
Acta mathematica see Mittag-Leffler, Gösta
algebra, history of 3, 33, 40–1
   see also Harriot, Thomas; Wallis, John
algorism 32
'Algus' 32
Arbuthnot, John, *Usefulness of Mathematical Learning* 44
arithmetic, history of 31–5, 37–40, 43–4

Baily, Francis, *Account of the Rev. John Flamsteed* 97, 101, 106
Barrow, Isaac 52, 54

Berlin Academy of Sciences 76
biography
   Freudian 67, 116
   mathematical 4, 71, 116–17
   scientific 5
   see also under individual biographers
Biot, Jean Baptiste 52, 60–1
Bjerknes, Carl Anton, biography of Abel 121–2, 126–30, 135
Bjørnson, Bjørnstjerne, poem for Abel's centennial 137
book trade
   English 11–12, 14
   French 75
Brewster, David 55–6, 97–8
   Life of Newton 61–2, 96
   Memoirs 63, 105–7
   see also De Morgan, Augustus
Buffon, Georges-Louis 81, 82, 88
Butler, William 37–9

Cajori, Florian 111, 154n
calculus
   fluxions 81–3, 89, 94, 111
   nature of 83–4, 86–7
   see also Leibniz, Gottfried Wilhelm; Newton, Isaac; Wallis, John
Church of England 5–6, 109, 112
Clark, Samuel 38–9, 41
Cocker, Edward, *Arithmetic* 42
Collins, John 11–14, 19
Crelle, August Leopold, *Journal* 119–20, 129
cycloid, history of 3, 24, 26–8

De Morgan, Augustus 5–6, 52, 58, 63–5
  alleged plagiarism by 113
  avoidance of national issues 91
  as biographer of Newton 90–1, 101–2, 105, 109–11, 114
  and Brewster 105–8, 110
  as historian 100
  and Libri 112–13
  life of 99–100
  and Newton–Leibniz dispute 97, 103–4, 108, 111–13
  nonconformism of 100, 109–10, 112–13
  and the Royal Society 103–4
  and UCL 100, 112–13
Descartes, René, alleged plagiarism from Harriot 15, 17, 24–5, 27, 151–3, 157
Donne, Benjamin 38, 42

Encyclopédie 5, 75–6, 87
Euclid 2, 31–2, 34–5, 54

Fenn, Joseph 40–1
Flamsteed, John *see* Baily, Francis
fluxions *see* calculus
Fontenelle, Bernard le Bovier de, *Éloge* for Newton 52–4, 56–7
founders, legendary 41–2

Galois, Évariste 118
'Geber' 33
genius 5, 51–8, 60–1, 67, 70, 90, 96, 99, 109, 117–18, 121, 129, 131, 137–8

Halley, Edmund 52
Harriot, Thomas
  algebra 149–54, 157, 162–3
  in America 145, 158–9
  Artis analyticae praxis 7, 15, 24, 146–50, 154, 160
  as astronomer 155–6
  Harriot seminar 160
  life and death 145–6, 158
  manuscripts
    difficulty of 154–6, 160, 163
    in the 17th century 146–9
    in the 18th and 19th centuries 154, 156–7
    in the 20th and 21st centuries 7, 157, 160–64
    see also Zach, Franz Xaver
  and Pell 148–9
  and Percy 145
  and publication 147
  recent work on 162–4
  speculation and error about 147, 158–60
  and Torporley 146, 158–9
  in Wallis's *Treatise of Algebra* 15, 17, 22
  and Warner 145
  see also Descartes, René; Lohne, Johannes; Robertson, Abram; Shirley, John; Stevens, Henry; Tanner, Cecily
hero, romantic 6, 118
Herodotus 35
historia 3, 20–3, 28–9
history
  Biblical 37–9, 41–2
  concept of 20–3, 28–9
  as cumulative 84–5, 88
  as progress/*progrès* 76, 78, 85, 88
  see also biography; *historia*; mathematics, history of ancient
Holmboe, Bernt Michael
  and Abel 130–31, 135, 142
  appointment of 133, 135–6
Holst, Elling, biography of Abel 122–4, 126, 131, 135–6
Hooke, Robert 14
Hutton, Charles 32, 34, 39
  Dictionary 31, 36, 153

Index

infinitesimals 74, 81–2, 88

Jacobi, C.G.J. *see* Abel, Niels Henrik, and Jacobi

Johnson, Samuel 33

Josephus 36, 38

Keill, John, and Newton–Leibniz dispute 94–5

Keynes, John Maynard 66

Kovalevskaya, Sofia *see* Mittag-Leffler, Gösta, and Kovalevskaya

Kronecker, Leopold *see* Mittag-Leffler, Gösta, and Kronecker

Laland, Joseph Jérôme de 5, 75

and Montucla's *Histoire* 74, 77, 79

Laplace, Pierre-Simon de 60

Leibniz, Gottfried Wilhelm 4

alleged plagiarism by 89, 93, 95–6, 105, 107–8, 113

character of 90, 95–6, 103, 108

discovery of calculus 92

Historia et origo 28–9

in London 93

and Wallis's *Treatise of Algebra* 17–18, 21

see also Newton, Isaac, dispute with Leibniz

Libri, Guglielmo *see* De Morgan, Augustus, and Libri

Lohne, Johannes, on Harriot 156, 162

Malcolm, Alexander 42–3

Arithmetic 33–7

Treatise of Music 34–5

mathematics

abstract/eternal 2, 38, 43, 51, 70, 82, 88

and commerce 31–4

editing historical 7–8

history of ancient 2–4, 14, 22–3, 31, 38–40, 42, 78

practical/useful 2, 39, 42, 43, 44, 50, 51, 54, 60–1, 74, 81, 88

see also algebra; arithmetic; calculus

Mittag-Leffler, Gösta

as Abel's heir 125–6, 137–44

and Abel's manuscripts 143

biography of Abel 115–16, 118, 124–6, 128–34, 136–9, 142–3

as editor of *Acta mathematica* 115, 119, 129, 140, 144

and Kovalevskaya 119

and Kronecker 118

life of 118, 125–6, 140–41

as mathematician 115, 139, 144

on Norwegian provincialism 132–3, 136

as outsider 126, 128, 130, 132–3, 136, 144

self-fashioning of 116, 120, 144

as teacher 115, 119, 134

and Weierstrass 118–19, 134, 140–41

monarchies, four 36

Montucla, Jean E.

avoidance of national issues 79–82, 88

Histoire des Mathématiques 5, 74, 76–88, 153

life of 75–7

Newton, Isaac 4, 5

alchemy of 63, 65–8

biographies of 47–9, 51–2, 55–7, 61, 69

character of 89–90, 96–9, 101–2, 108–10, 114

Commercium epistolicum 28, 53, 80–3, 85, 95, 102–3, 105–6, 111

discovery of calculus 91–2

dispute with Leibniz 4–5, 28, 48, 53, 63, 69, 73–4, 79–85, 87, 89–91, 93–7, 101–2, 105–8, 111, 114

in Wallis's *Treatise of Algebra* 18, 25

as an experimenter 49–50
language of 48, 51, 54, 56–9, 67
Mathematical Papers 65
as a mathematician 47–71
optics of 61–2
Principia mathematica 48, 50, 54, 56–7, 59, 62, 67, 80, 97
see also De Morgan, Augustus; Fontenelle, Bernard le Bovier de; Brewster, David; calculus; Keill, John; Keynes, John Maynard; Leibniz, Gottfried Wilhelm; Rigaud, Stephen Peter; Royal Society; Westfall, Richard; Whewell, William
Nile, Flooding of 37
Norway *see* Abel, Niels Henrik; Mittag-Leffler, Gösta; patriotism, Norwegian
numerals, Hindu–Arabic 32, 34

patriotism, Norwegian 124, 126–7, 135–7, 143–4
pedagogy 40–1
Pell, John *see* Harriot, Thomas, and Pell
Pepys, Samuel 9
Percy, Henry *see* Harriot, Thomas, and Percy
Phoenicians 31, 34, 36, 39, 42
Playford, John 14, 16
Priestley, Joseph 57
Proclus 35

Ramus, Peter, *Prooemium mathematicum* 2
readers
popular 3, 4, 8, 51, 57, 62–3, 66, 70
young 40–2
Rigaud, Stephen Peter
on Harriot 156
on Newton 98

Robertson, Abram, and Harriot's manuscripts 155–6
Rouse Ball, W.W. 65, 100n, 104n, 111
Royal Society 6, 17
and papers of Jeremiah Horrox 21–2
and Wallis's *Treatise of Algebra* 11, 13–14, 16, 18–19, 21
see also De Morgan, Augustus, and the Royal Society; Harriot, Thomas, manuscripts, in the 17th century; Newton, Isaac, dispute with Leibniz

Savile, Henry
lectures on Ptolemy 1
Praelectiones tresdecim 2
series 16, 19, 25, 29, 80–1, 83, 85–6, 91–3
Seth, pillars of 36
Shirley, John, *Thomas Harriot* 161
Smith, David E. 111
Stevens, Henry, *Thomas Harriot and his associates* 157–9
Strabo 35

Tanner, Cecily, on Harriot 160
Torporley, Nathaniel *see* Harriot, Thomas, and Torporley

Vossius, Gerardus 21–2

Wallis, John 9, 13
and calculus 74, 83–5, 93
as decoder 9
and Harriot's ideas 149–51
Opera mathematica 9, 11
portrait by Kneller 9, 10, 29–30
A Proposal about Printing a Treatise of Algebra 13–15, 17
as Savilian professor 9, 12, 17
Treatise of Algebra 3, 7, 9–11, 14–26, 28–30, 148–53

publication of 12–14, 16, 19
see also Harriot, Thomas; Leibniz,
Gottfried Wilhelm; cycloid,
history of
Warner, Walter see Harriot, Thomas, and
Warner
Weierstrass, Karl see Abel, Niels Henrik,
and Weierstrass; Mittag-Leffler,
Gösta, and Weierstrass

Westfall, Richard 47, 66–8
Whewell, William 52
History of the Inductive Sciences 99
on Newton 108–9
Whiston, William 36

Zach, Franz Xaver, and Harriot's
manuscripts 154–5

831116